Nelson MindTap + **You** = Learning amplified

"I love that everything is interconnected, relevant and that there is a clear learning sequence. I have the tools to create a learning experience that meets the needs of all my students and can easily see how they're progressing."

— **Sarah,** Secondary School Teacher

Nelson Science Year 9 Western Australia Student Book
1st Edition
Mya Skirving
Anne Disney
Florence Coghlan
Celia McNeilly
ISBN 9780170472838

Series publisher: Catherine Healy
Project editor: Alan Stewart
Editor: Catherine Greenwood
Series text design: Leigh Ashforth
Series cover design: Leigh Ashforth
Series designer: Linda Davidson
Cover image: Shutterstock.com/Priyank Dhami
Permissions researcher: Debbie Gallagher, Brendan Gallagher
Production controller: Bradley Smith
Typeset by: MPS Limited

Any URLs contained in this publication were checked for currency during the
production process. Note, however, that the publisher cannot vouch for the
ongoing currency of URLs.

© 2023 Cengage Learning Australia Pty Limited

For product information and technology assistance,
in Australia call **1300 790 853**;
in New Zealand call **0800 449 725**

For permission to use material from this text or product, please email
aust.permissions@cengage.com

National Library of Australia Cataloguing-in-Publication Data
A catalogue record for this book is available from the National Library of
Australia.

Cengage Learning Australia
Level 5, 80 Dorcas Street
Southbank VIC 3006 Australia

Cengage Learning New Zealand
Unit 4B Rosedale Office Park
331 Rosedale Road, Albany, North Shore 0632, NZ

For learning solutions, visit **cengage.com.au**

Printed in China by 1010 Printing International Limited.
1 2 3 4 5 6 7 27 26 25 24 23

nelson science.

9

Mya Skirving
Anne Disney
Florence Coghlan
Celia McNeilly

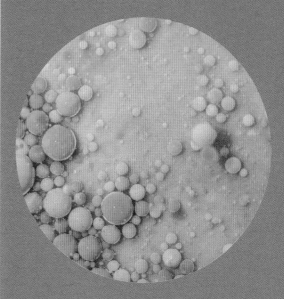

This cover image shows a mixture of milk, oil and acrylic paint. This type of mixture is known as an emulsion. Liquids in emulsions are usually immiscible, meaning they cannot mix with each other.

Milk itself is an emulsion, with droplets of oil dispersed in water. Oil is classified as 'hydrophobic' because it is repelled by water molecules. Some of the droplets you can see are the water-based acrylic paint suspended in oil, while others are droplets of oil suspended in the water component of the milk.

WA
Australian Curriculum

FIRST NATIONS AUSTRALIANS GLOSSARY

Country/Place

Spaces mapped out that individuals or groups of First Nations Peoples of Australia occupy and regard as their own and that have varying degrees of spirituality. These spaces include lands, waters and sky.

Cultural narrative

A broad term that encompasses any cultural expression that includes (but is not limited to) knowledge and community values that are central to the identity of a particular group of First Nations Peoples.

Cultural narratives can hold information about almost anything, such as the origins of life, or can teach people about acceptable behaviour and rules, such as caring for Country.

They can take the form of songs, stories, visual arts or performances. 'Cultural narrative' is a more accurate and respectful term than 'myth', 'story' or 'fable'; terms that often diminish their importance.

First Nations Australians

'First' refers to the many nations/cultures who were in Australia before British colonisation. This a collective term that refers to all Aboriginal Peoples and Torres Strait Islander Peoples. The term 'Indigenous Australians' is also used to refer to First Nations Australians.

Nation

A self-governed community of people based on a common language, culture and territory.

Peoples and Nations

We use the plural for these terms because First Nations Australians do not belong to one nation/culture. There are many distinct Peoples and Nations. Also, some Nations consist of distinct clans or groups, so are referred to as Peoples.

ACKNOWLEDGEMENT OF COUNTRY

Nelson acknowledges the Traditional Owners and Custodians of the lands of all First Nations Peoples of Australia. We pay respect to their Elders past and present.

We recognise the continuing connection of First Nations Peoples to the land, air and waters, and thank them for protecting these lands, waters and ecosystems since time immemorial.

Contents

Authors and contributors

Lead author

Mya Skirving

Contributing authors

Anne Disney

Florence Coghlan

Celia McNeilly

Consultants

Joe Sambono

First Nations
curriculum consultant

Dr Silvia Rudmann

Digital learning
consultant

Judy Douglas

Literacy
consultant

Science communication
consultants

Additional science
investigations

Reviewers

**Anne McGregor, Scott
Adamson, Megan Mackay,
Faye Paioff, Pete Byrne**

Teacher reviewers

**Aunty Gail Barrow, Nicole Brown, Christopher Evers,
Associate Professor Melitta Hogarth, Carly Jia,
Jesse King, Dr Jessa Rogers, Theresa Sainty**

Reviewers of First Nations Science Contexts pages

9780170472838

nelson science. Learning Ecosystem

Nelson Science 9 caters to all learners

Nelson Cengage has developed a **Science Learning Progression Framework**, which is the foundation for Nelson's Science 7–10 series. An editable version is available on Nelson MindTap.

Reinforce
Nelson MindTap provides a wealth of differentiated activities and resources to meet the needs of all students.

LEARN

Evaluate prior knowledge
Students complete a quiz to test their prior knowledge.

LEARN

nelson science. Nelson MindTap

Engage
Each chapter showcases fascinating, real-world science in action, while our hands-on activities, short videos and fun interactives keep students engaged.

LEARN

Assess
Allocate and grade assessments using our differentiated end-of-topic tests and summative portfolio assessment tasks in Nelson MindTap.

LEARN

Practise
Our differentiated, scaffolded activities and investigations allow all learners to build essential skills and knowledge.

LEARN

 **Nelson MindTap**

A flexible and easy-to-use online learning space that provides students with engaging, tailored learning experiences.

- Includes an eText with integrated activities and online assessments.
- Margin links in the student book signpost multimedia student resources found on Nelson MindTap.

Video activity
Cells

For students:

- Short, engaging videos with fun quizzes that bring science to life.
- Interactive activities, simulations and animations that help you develop your science skills and knowledge.
- Content, feedback and support that you can access as you need it, which allows you to take control of your own learning.

For teachers:

- 100% modular, flexible courses let you adapt the content to your students' needs.
- Differentiated activities and assessments can be assigned directly to the student, or the whole class.
- You can monitor progress using assessment tools like Gradebook and Reports.
- Integrate content and assessments directly within your school's LMS.

How to use this book

Big science, real context: The opening page begins the chapter by placing the science topic into a real-life context that is both interesting and relevant to students' lives.

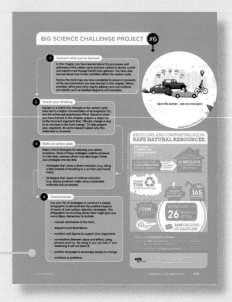

Think, do, communicate: You are encouraged to reflect on and apply your learning to a set of activities, which allows you to make meaningful connections with the content and skills you have just learned.

Learning modules: Content is chunked into key concepts for effective teaching and learning.

Learning objectives: Clear, concise objectives give you oversight of what you are learning and set you up for success.

Key words: These are defined the first time they appear.

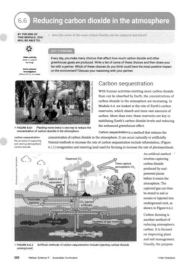

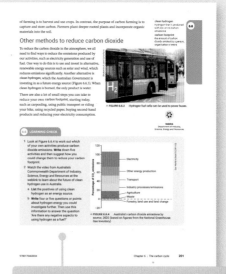

Learning check: These are engaging activities to check your understanding. Activities are presented in order of increasing complexity to help you confidently achieve the module's learning objectives. **Bolded** cognitive verbs help you clearly identify what is required of you. Activities are presented in order of increasing complexity.

Science as a Human Endeavour: Elaborations are explicitly addressed with interesting, contemporary content and activities.

First Nations science contexts: This content was developed in consultation with a First Nations Australian curriculum specialist. It showcases the key Aboriginal and Torres Strait Islander History and Cultural Elaborations, with authentic, engaging and culturally appropriate science content.

Activities: Activities are open-ended and often hands on, helping you understand the connections between First Nations cultures and histories and science.

Science skills in focus: Each chapter focuses on a specific science investigation skill. This is explained and modelled with our Science skills in a minute animation, before you put it into practice in a science investigation. The science skill is reinforced with our Science skills in practice digital activities.

Investigations: Practise and reinforce good scientific method through fit-for-purpose, hands-on science investigations.

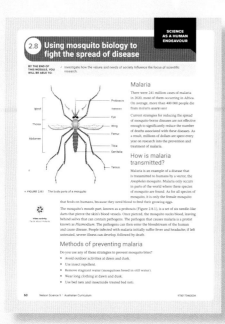

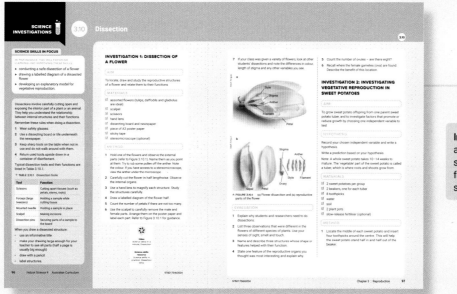

1

Homeostasis

9780170472838

iStock.com/AleksandarNakic

▲ **FIGURE 1.0.1** Sweating is one way our bodies maintain our body temperature.

Do you get really hot in summer or during exercise? In 2021, people living in Julien Creek in Queensland experienced a 43.6°C day, while people living in the town of Wyndham in Western Australia experienced temperatures of 45°C. Air conditioning and sweating are two different ways of cooling down. Air conditioning is expensive, and generally not good for the environment. Sweating is our body's natural mechanism for cooling down; however, excessive sweating can be life-threatening due to dehydration.

▶ What alternative, effective strategies can we choose instead of air conditioning to help us stay cool on a hot day?

#1 SCIENCE CHALLENGE ACCEPTED!

At the end of this chapter, you can complete the Big Science Challenge Project #1. You can use the information you learn in this chapter to complete the project.

Assessments
- Prior knowledge quiz
- Chapter review questions
- End-of-chapter test
- Portfolio assessment task: Science investigation

Videos
- Science skills in a minute: Using a microscope **(1.9)**
- Videos activities: Nervous system **(1.2)**; Neurons as cells **(1.3)**; The brain **(1.4)**; Endocrine system **(1.5)**; What is diabetes? **(1.8)**

Science skills resources
- Science skills in practice: Using a microscope to examine cells **(1.9)**
- Extra science investigations: Negative feedback loops **(1.1)**; Dissecting a spinal cord **(1.2)**

Interactive resources
- Label: Negative feedback model **(1.1)**; Parts of the brain **(1.4)**; Endocrine system **(1.5)**; Negative feedback loops **(1.7)**
- Drag and drop: Different types of neurons **(1.3)**; Maintaining body temperature **(1.6)**

✷ Nelson MindTap

To access these resources and many more, visit:
cengage.com.au/nelsonmindtap

Interactive resources
Label: Negative feedback model

Extra science investigation
Negative feedback loops

GET THINKING

Have you heard of the term 'homeostasis'? As you scan this first module, try to find some familiar scientific terms and take note of the terms you do not know yet.

Homeostasis is a balancing act

On hot days setting your air conditioner to a specific temperature keeps your house at a constant, comfortable temperature. When the temperature of the room goes above or below the set temperature, the air conditioner's sensors detect the change, and the air conditioner switches on or off. Similarly, when our body's temperature rises above or below 37°C, processes inside our body 'switch on' or 'switch off' to regulate our body's temperature and keep it stable. This balancing act is known as **homeostasis**.

Tolerance range and homeostasis

Your body responds to changes in internal or external conditions, such as temperature, blood glucose levels, water and salts, and uses different processes to keep these factors within the specific ranges you can tolerate. This known as a **tolerance range**. If the conditions inside your body change too far from these ranges, you can become sick or even die (Figure 1.1.1).

homeostasis
the maintenance of a constant internal environment necessary for survival

tolerance range
the range of a particular condition inside the body that an organism can survive

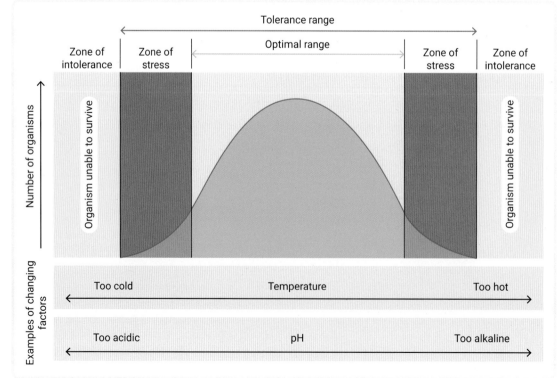

► **FIGURE 1.1.1** For survival, internal factors need to be maintained within a tolerance range. The body works best when these factors are within optimal range.

Our bodies function best when factors are within an optimal range. Beyond this range, our bodies become stressed, but we can usually survive unless the factor deviates into the zone of intolerance. Different organisms have different ranges within which they can survive.

Have you done the ice bucket challenge? Many celebrities have, to fundraise for charity. The challenge involves pouring a bucket of ice water over your head, as shown in Figure 1.1.2. The cold water causes the internal temperature to decrease. The change in temperature is a stimulus that is detected by the body. The body responds by shivering and producing goose bumps. In 2021, one celebrity's internal body temperature dropped until her shivering turned into violent shaking.

Some athletes use an ice bath or very cold water to help muscle recovery (Figure 1.1.3). They follow strict instructions about how long they can stay in the cold water so that their body temperature does not fall below the tolerance range.

While the athletes are in cold water, their internal body temperature astonishingly remains at 37°C. This homeostasis is possible through mechanisms such as shivering.

▲ **FIGURE 1.1.2**   The ice bucket challenge

▲ **FIGURE 1.1.3**   Athletes recovering in an ice bath

Stimulus–response models

Scientists have developed a model called the stimulus–response model to help us understand the mechanisms that work to maintain our body's constant internal environment (see Figure 1.1.4).

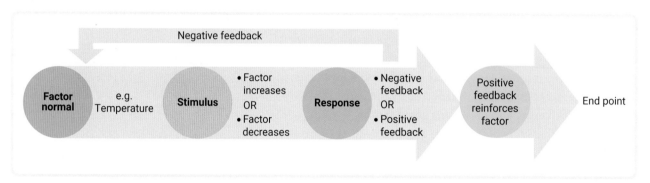

▲ **FIGURE 1.1.4**　The stimulus–response model

Stimulus: detecting change

stimulus
a change in a factor above or below the optimal range; causes a movement away from normal

receptor
a specialised cell that detects a stimulus; may be internal or external

A **stimulus** is a mechanism that causes a change in the environment (inside or outside the body) that can be detected by one of the body's **receptors**. For example, an increase or decrease in environmental temperature (e.g. a cold wind) is a stimulus. A receptor is any part of the body (cell, tissue or organ) that can receive information from a stimulus. For example, thermoreceptors in the skin and brain are receptors for temperature. A receptor detects a stimulus when conditions go above or below its normal narrow range. So, when your body is cold in winter or hot in summer, receptors will detect these changes.

Response: reacting to change

coordinating centre
an organ or tissue that receives and processes information from receptor cells and coordinates a response

response
the action of an effector that reverses a change in condition due to detection of a stimulus

effector
a muscle or gland that receives a message from the coordinating centre and carries out a response

A **coordinating centre** receives information from the receptor, processes the information and coordinates a **response**. For example, the hypothalamus is the part of the brain that acts as the coordinating centre for temperature regulation. The coordinating centre sends signals to an **effector** to carry out a response. An effector is any muscle or gland that does the work to return the body's condition to normal. For example, when your body is too cold, the hypothalamus sends a message to muscles attached to your hair follicles, telling them to contract. This makes your hairs stand up and conserves heat. You may know this as 'goose bumps'.

Returning to normal range

negative feedback
a mechanism in which a response reduces a condition or factor back to its optimal range

positive feedback
a mechanism in which a response reinforces the condition or factor until an end point

A response is the action the body takes to return conditions inside the body to the normal range. When the response reduces the condition or factor back to its optimal range, the mechanism is known as **negative feedback**. When the condition or factor is further increased away from normal, the mechanism is called **positive feedback**.

Negative feedback

When the body detects a stimulus, it will act to restore normal conditions. The effects of the stimulus can be reduced through a negative feedback mechanism.

9780170472838

We use the negative feedback model to understand the steps involved in homeostasis. The model represents homeostasis as a circuit, which suggests this process occurs continuously inside our bodies. However, in reality it is a bit like the air conditioner that switches on and off to bring the room temperature back to the set temperature.

The negative feedback mechanism can be modelled by a flow diagram (Figure 1.1.5).

Positive feedback

If a response forces a stimulus to increase instead of returning to normal, the mechanism is referred to as positive feedback. Positive feedback is rare. One example is blood clotting. When skin is cut, special white blood cell fragments, known as platelets, release clotting factors. This causes more platelets to be transported to the wound to help form a clot, preventing blood loss and forming a scab. Positive feedback continues until a certain point has been reached, such as the finished formation of a blood clot.

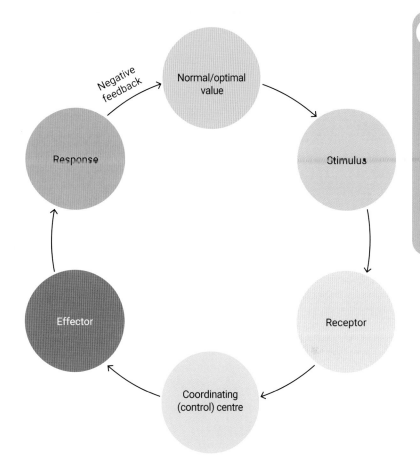

▲ **FIGURE 1.1.5**   A negative feedback model

1.1 LEARNING CHECK

1 **Define** 'homeostasis'.
2 **Describe** negative feedback.
3 Find a friend who currently has a wound. Does it have a scab? **Describe** the feedback mechanism their body went through to form it.
4 **Compare** and **explain** the principles of negative and positive feedback mechanisms by copying and completing the following table.

Principle	Negative feedback	Positive feedback
Role in homeostasis and the survival of an organism		
Deviation of factor (e.g. temperature) from normal		
Frequency (often or rare)		

GET THINKING

Scan the information in this module and see which words best describe the structures and functions of the nervous system.

central nervous system (CNS)
the brain and spinal cord

nerve
a collection of fibres, surrounded by a protective coat, that transmit messages as nerve impulses to and from the CNS

Parts of the nervous system

Animal systems depend on the nervous system as a communication network. Specific functions of the nervous system include monitoring change, transmitting messages and coordinating responses.

The nervous system consists of two parts: the central nervous system and the peripheral nervous system, as shown in Figure 1.2.1.

Central nervous system

The **central nervous system (CNS)** consists of the brain and spinal cord. Its function is to detect a stimulus, interpret and process information, and coordinate a response. The CNS is connected by **nerves** to receptors in tissue where our five senses are at work: sight, hearing, taste, touch and smell. When the brain or spine receives information from the receptors, they coordinate a response. Responses vary enormously and include avoiding danger, resetting body temperature and readjusting our eyes to manage bright light.

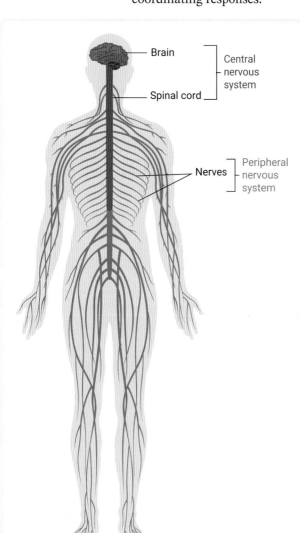

▲ **FIGURE 1.2.1**   The two parts of the nervous system: the central nervous system (red) and peripheral nervous system (blue)

9780170472838

Peripheral nervous system

The **peripheral nervous system (PNS)** is made up of a network of nerves. The nerves are made of **nerve fibres** that transmit messages in the form of **nerve impulses** to and from the CNS.

Messages travel along nerves in the PNS. Each nerve is a collection of nerve fibres, as shown in Figure 1.2.2. Nerves extend from the spinal cord to all muscles, glands and organs in the body.

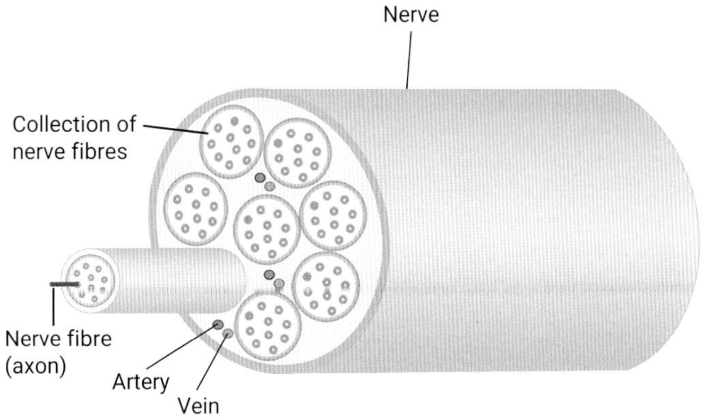

The PNS consists of the sensory division and the motor division, which can be further divided into two parts: the autonomic nervous system and the somatic nervous system, as shown in Figure 1.2.3. Both systems transport nerve impulses away from the CNS to muscles or glands (effectors) to carry out a response.

The autonomic nervous system performs involuntary functions that we are not conscious of, such as sweating, shivering, the beating of our heart and insulin release.

The somatic nervous system performs voluntary functions that we are conscious of, such as muscular movements required to walk, eat and game.

▲ **FIGURE 1.2.2** The structure of a peripheral nerve in the human body

peripheral nervous system (PNS)
the network of nerves outside the brain and spinal cord

nerve fibre
the section of a nerve cell that carries nerve impulses away from the cell body

nerve impulse
an electrical message that is transmitted along nerves to and from the CNS

Video activity
Nervous system

Extra science investigation
Dissecting a spinal cord

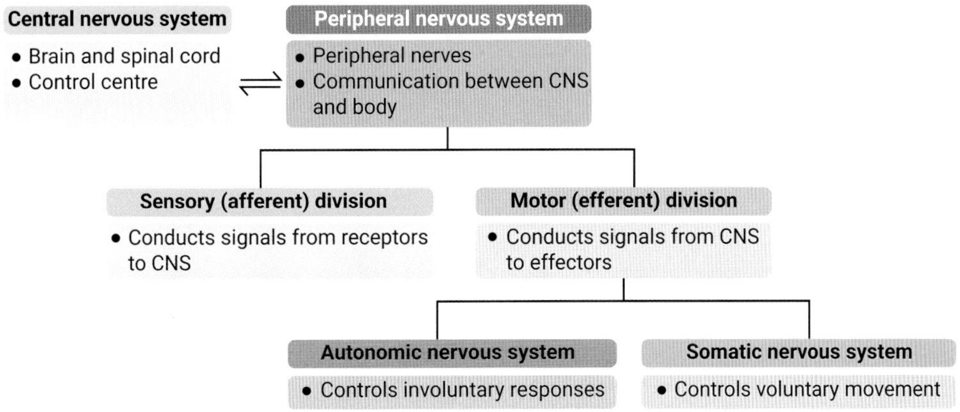

Central nervous system
- Brain and spinal cord
- Control centre

⇌

Peripheral nervous system
- Peripheral nerves
- Communication between CNS and body

Sensory (afferent) division
- Conducts signals from receptors to CNS

Motor (efferent) division
- Conducts signals from CNS to effectors

Autonomic nervous system
- Controls involuntary responses

Somatic nervous system
- Controls voluntary movement

▲ **FIGURE 1.2.3** The divisions of the nervous system

1.2 LEARNING CHECK

1 **State** the main parts of the CNS and the PNS.
2 **Draw** a diagram showing the distinct parts of the CNS and the PNS.
3 **Describe** the function of the autonomic part of the PNS.
4 **Research** the sensory division and the motor division of the PNS. Write 3–5 sentences about each function and include examples.

Video activity
Neurons as cells

Interactive resource
Drag and drop:
Different types of
neurons

neuron
a specialised cell that can transmit nerve impulses; also known as a nerve cell

dendrite
a branching network at the end of a neuron that receives information from other neurons

cell body
the part of a neuron that contains the cytoplasm, including the nucleus

axon
a long, thin fibre that carries electrical impulses from the cell body of a neuron towards the next neuron

myelin sheath
a protective coat around an axon that increases the speed of nerve impulses

GET THINKING

Prepare a concept map to summarise the content in Modules 1.1 and 1.2. As you work through Modules 1.3–1.5, add new the information to your concept map. You may wish to include diagrams to help your understanding.

Neurons

Information is carried around the human body at mesmerising speeds, up to 120 metres per second ($m\,s^{-1}$). The structures that carry the information are highly specialised nerve cells – **neurons**. There are an estimated 100 billion neurons in the brain and spinal cord combined, and many more specialised neurons in the peripheral nervous system.

As shown in Figure 1.3.1, a neuron has branching, finger-like **dendrites** at one end of its cell body that receive information from other cells in the form of chemical signals. The **cell body** contains the nucleus. A long, thin fibre, an **axon**, extends from the other end of the cell body. Electrical signals flow along the axon towards the next neuron. The **myelin sheath** is the fatty, insulating layer around axons. The functions of each part of a neuron are described in Table 1.3.1.

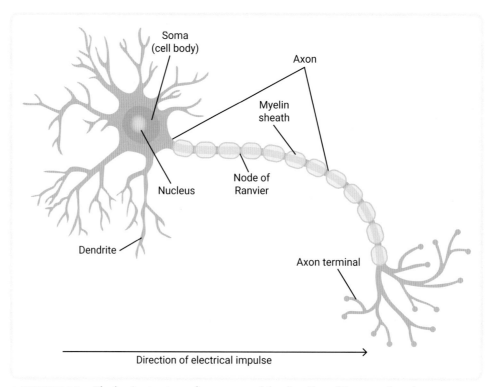

Soma (cell body)
Axon
Myelin sheath
Nucleus
Node of Ranvier
Dendrite
Axon terminal
Direction of electrical impulse

▲ **FIGURE 1.3.1**   The basic structure of a neuron and the direction of the nerve impulse

9780170472838

▼ TABLE 1.3.1 The basic neuron structures and their functions

Structure	Functions
Dendrite	Receives information from other neurons in the form of chemical signals
Cell body	Contains the nucleus
Axon	Conducts (carries) electrical impulses from the cell body, or another cell's dendrites, to the end of the neuron (axon terminals)
Myelin sheath	Increases the speed of nerve impulses
	Insulating (high electrical resistance) fatty layer that protects the nerve fibres from other electrical signals

Neuron communication

A neuron conducts an electrical signal along its axon, as shown in Figure 1.3.2. In some cases, the signal needs to transform into a chemical signal to flow across the gap (**synapse**) between most neurons. If there were no gaps, the network of neurons would conduct the electrical impulses continuously and uncontrollably. Imagine your home without light switches. You would not have any control over switching the lights off and on.

synapse
the gap between two neurons

Synapses provide some communication control. When an electrical impulse arrives at the axon terminal, it activates the release of a **neurotransmitter**. The neurotransmitter delivers the message across the synapse to the dendrite of the next neuron, or a muscle or gland cell. At the synapse, the impulse can be modified to be made bigger or smaller depending on the needs of the body. Once the message has been delivered across the synapse, the message is converted back into an electrical impulse.

neurotransmitter
a chemical signal that delivers a message that was in the form of an electrical impulse

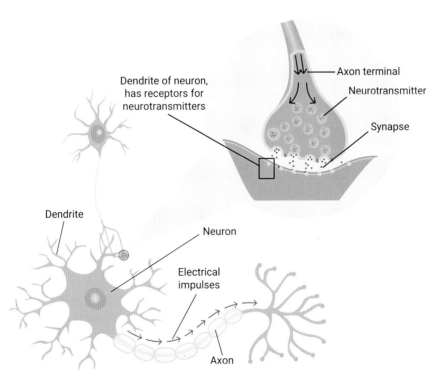

Dendrite of neuron, has receptors for neurotransmitters

Axon terminal

Neurotransmitter

Synapse

Dendrite

Neuron

Electrical impulses

Axon

▲ FIGURE 1.3.2   Neurotransmitters pass messages across a synapse from the axon terminal of one neuron to the receptors on the dendrite of the next neuron.

Three types of neurons

Figure 1.3.3 shows the structure and function of the three main types of neurons – sensory neurons, motor neurons and interneurons. The function of each type is possible because of their unique structure. They can be classified according to the direction in which they send nerve impulses.

- **Sensory neurons** have sensory structures that detect change in their environment. They send an electrical impulse with this information towards the CNS (afferent direction).
- **Interneurons** are shorter neurons that have many branching dendrites to communicate a lot of information between sensory neurons and motor neurons.
- **Motor neurons** have long axons reaching effector muscles or glands with information sent from the CNS (efferent direction).

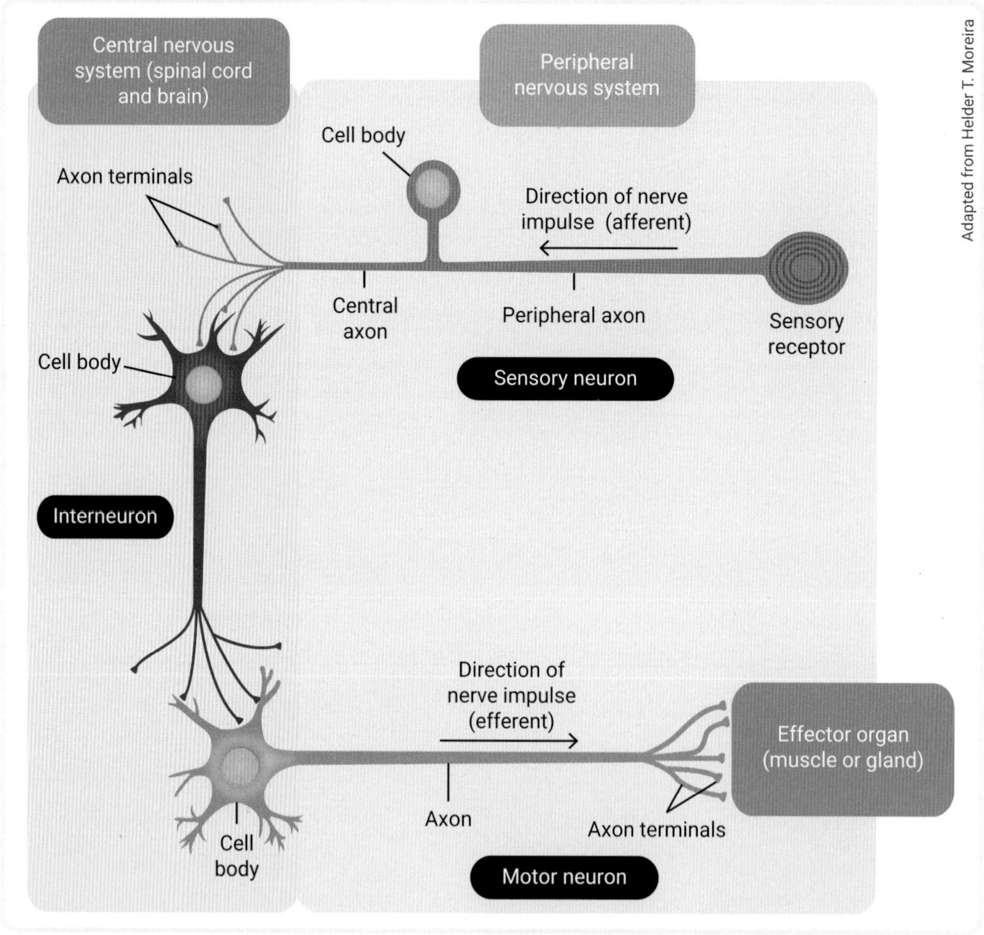

sensory neuron
a neuron that sends electrical impulses from receptors to the CNS

interneuron
a short neuron that sends nerve impulses between sensory and motor neurons within the CNS

motor neuron
a neuron that sends electrical impulses from the CNS to effectors

Adapted from Helder T. Moreira

▲ **FIGURE 1.3.3** The three types of neurons work together to send nerve impulses towards or away from the CNS.

The structures and functions of these three types of neurons are summarised in Table 1.3.2.

Sometimes the myelin sheath is damaged. This can be the result of a disease known as multiple sclerosis – a disease that results in nerve damage and causes symptoms such as weakness, fatigue and blurry vision.

9780170472838

▼ TABLE 1.3.2 The structure and function of the three main types of neurons

	Sensory neuron	Interneuron	Motor neuron
Structure	Receptors at one end of a relatively long axon with a cell body branching off the axon closer to the axon terminals	Relatively short but branching dendrites, with a cell body and branching axons	Many branching dendrites at one end of a large cell body with a long axon ending with axon terminals
Location	Sensory organs such as eyes and skin, in PNS	CNS, within brain and spinal cord	Dendrites in CNS, axon terminals in PNS
Function	To detect changes and transmit nerve impulses towards spinal cord or brain	To transmit nerve impulses from a sensory neuron or other interneurons to a motor neuron or other interneurons	To transmit nerve impulses from brain and spinal cord to a muscle or gland to carry out a response
Direction of impulse	From sensory receptors to CNS (afferent direction)	Between sensory and motor neurons	From CNS to an effector muscle or a gland (efferent direction)

1.3 LEARNING CHECK

1 Visit the Queensland Brain Institute website to view various stained microscope images of neurons. **Observe** and **describe** a typical neuron.

2 **Define** 'neurotransmitter'.

3 Three common neurotransmitters are acetylcholine, serotonin and dopamine. Choose one to **research**. Write five sentences about its function.

4 **Explain** why the axon part of a neuron is covered in myelin.

5 **Investigate** the speed of your nerve impulses (reaction time). Ask a partner to hold a ruler between your thumb and first finger (don't grip the ruler). Line up 0 cm at your finger. Your partner drops the ruler while you try to catch it. Record the length on the ruler where you caught it (Figure 1.3.4). Swap and repeat. The person with the shortest length may have the fastest impulses!

Weblink
Queensland Brain Institute

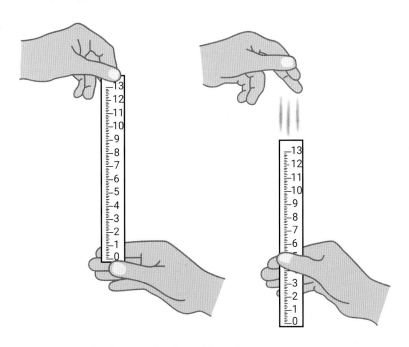

▲ FIGURE 1.3.4 Testing reaction time with a ruler

BY THE END OF THIS MODULE, YOU WILL BE ABLE TO:
- ✓ describe the structures and functions of the brain and its parts: cerebrum, cerebellum, pituitary gland, hypothalamus and brain stem
- ✓ explain how changes to the brain can affect body functions.

Video activity
The brain

Interactive resource
Label: Parts of the brain

GET THINKING

How much do you know about the human brain? After you read this module, reflect on which part of the brain fascinates you the most.

Brain facts

What memories do you have of last school holidays? Where are those memories filed away? They are stored in a very complex organ, the **brain**. Situated within the protective bones of the cranium (skull), the brain is about 1.4 kg of living tissue that does your thinking, learning, feeling and remembering. The brain is the coordinating centre of the CNS.

The brain contains grey matter (mostly neuron cell bodies) and white matter (mostly bundles of axons found in the interior part of the brain). Nerve impulses between neurons are generated when the brain is performing its major functions such as learning science, feeling emotions, monitoring and analysing sensory information, and coordinating a response.

The brain has a symmetrical structure: it has a right **hemisphere** and a left hemisphere. Most of the knowledge we have today about the brain was gained through technological advancements in brain imaging such as magnetic resonance imaging (MRI). The left hemisphere controls motor function on the right side of the body and the right hemisphere of the brain controls motor function on the left side of the body. The image in Figure 1.4.1a was made possible because of MRI.

brain
a very complex organ, the coordinating centre of the CNS

hemisphere
one of the two symmetrical halves of the brain

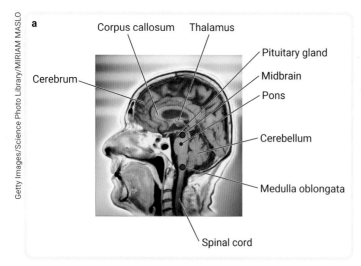
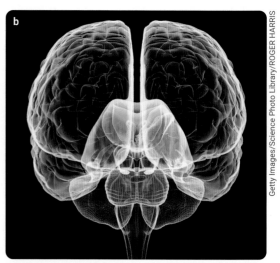

▲ **FIGURE 1.4.1** **(a)** An MRI image of one hemisphere of the brain. The brain stem consists of the midbrain, pons and medulla oblongata (marked in red). **(b)** The symmetrical structure of the brain, the hemispheres.

External brain anatomy

The largest part of the brain, the **cerebrum**, is above the **cerebellum** and **brain stem**. Major functions of the cerebrum include controlling movement, learning, emotion, memory and perception. The wrinkly, grey, outer layer is the cerebral cortex, the information processing centre of the brain. Scientists studying the brain have found that four sections of the cerebrum, called lobes, carry out distinct functions, as shown in Figure 1.4.2.

The two hemispheres communicate via the **corpus callosum**, which is a collection of axons that extend across the two hemispheres. The corpus callosum is near the centre of the brain. It allows comparison and combination of the sensory inputs from the left and right sides of the body.

The cerebellum is tucked under the cerebrum and behind the brain stem. When the cerebellum is damaged or old, people become uncoordinated and can lose their balance. This indicates that some of the functions of the cerebellum include coordination when moving, and balance; for example, when walking.

The brain stem consists of the midbrain, the pons and the medulla oblongata.

- The midbrain receives sensory information, such as sight and sound, and transfers it to the cerebrum.
- The pons transmits messages between the PNS and midbrain.
- The medulla controls automatic (and involuntary) activities such as breathing, heart rate, digestion and vomiting.

cerebrum
the largest part of the brain; controls movement, learning, emotion, memory and perception

cerebellum
a smaller section at the back of the brain; controls movement and balance

brain stem
consists of the midbrain, the pons and the medulla oblongata; controls automatic and involuntary activities such as breathing, heart rate, digestion and vomiting

corpus callosum
a collection of axons that extend across the two hemispheres of the brain that compares and combines the sensory inputs from the left and right sides of the body

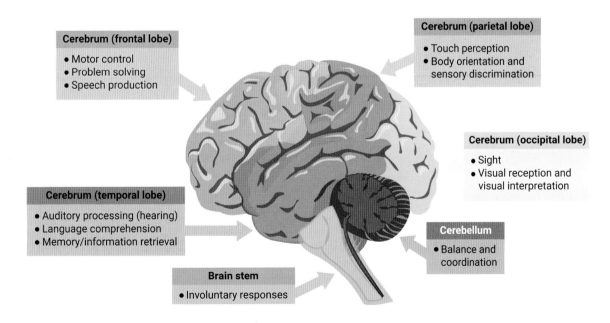

Cerebrum (frontal lobe)
- Motor control
- Problem solving
- Speech production

Cerebrum (parietal lobe)
- Touch perception
- Body orientation and sensory discrimination

Cerebrum (occipital lobe)
- Sight
- Visual reception and visual interpretation

Cerebrum (temporal lobe)
- Auditory processing (hearing)
- Language comprehension
- Memory/information retrieval

Cerebellum
- Balance and coordination

Brain stem
- Involuntary responses

▲ **FIGURE 1.4.2**   The external anatomy of the brain

Internal brain anatomy

hypothalamus
a pea-sized structure of the brain; plays a major role in homeostasis

pituitary gland
a gland that produces and secretes several hormones; is connected to, and is controlled by, the hypothalamus

The **hypothalamus** works with the **pituitary gland** to play a major role in homeostasis. The hypothalamus (*hypo* = low) is located below the cerebrum and in front (anterior) of the brain stem. Although it is only the size of a pea, it has important functions. Major functions of the hypothalamus include:

- detecting changes in temperature and water levels within the body
- coordinating negative feedback responses
- regulating hunger and thirst
- controlling the fight-or-flight response
- producing hormones
- controlling the pituitary gland.

Feeling thirsty? The hypothalamus produces antidiuretic hormone (ADH) and stores it in the pituitary gland. Your pituitary gland can release ADH when you need to conserve water. The pituitary gland can produce and secrete many other hormones, such as growth hormones, which are essential for homeostasis and sexual development. The pituitary gland sits just under and is connected to the hypothalamus.

A traumatic brain injury can disrupt the functioning of the hypothalamus and the pituitary gland. Signs that the hypothalamus and the pituitary gland have been injured or disrupted include reduced growth and weight, and unregulated water levels and body temperature. These symptoms can be attributed to hormone deficiencies resulting from injuries to certain structures.

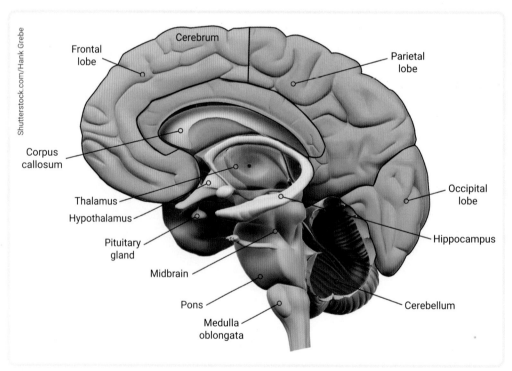

▲ **FIGURE 1.4.3**   The internal anatomy of the brain

9780170472838

1 **Identify** the part of the brain that controls several homeostasis processes in the body.

2 **Describe** the structure of the brain in two or three sentences.

3 **Summarise** the functions of different parts of the brain by copying and completing the following table.

Part of the brain	Function
Corpus callosum	
Brain stem	
Pituitary gland	
Cerebellum	
Hypothalamus	

4 A student recently suffered a traumatic brain injury during a car accident. One of his symptoms is that he is excessively thirsty all the time. What information about the brain could you share with the student that may help him understand the cause of this disruption to water homeostasis? **Research** traumatic brain injury to find other symptoms.

5 Copy the following diagram and **label** the main parts of the brain's external anatomy.

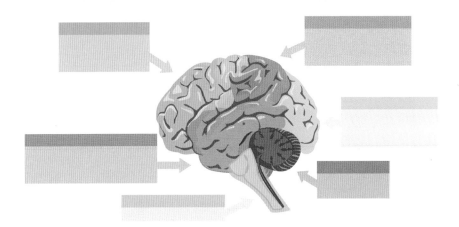

6 **Draw** the internal brain anatomy as shown in Figure 1.4.3. Try to label the sketch from memory.

1.5 The endocrine system

BY THE END OF THIS MODULE, YOU WILL BE ABLE TO:

✓ describe the endocrine system, including the main glands and their major hormones and functions

✓ describe, explain and compare the functions of different types of hormones

✓ compare the endocrine and nervous systems and demonstrate their interdependence.

Video activity
Endocrine system

Interactive resource
Label: Endocrine system

endocrine system
a network of glands that secrete hormones into the bloodstream to be transported to target cells

gland
a tissue that releases hormones

hormone
a chemical messenger

GET THINKING

Return to the concept map you started at the beginning of Module 1.3. Do you now have more information to add to the map? Scan this module to preview which vocabulary and concepts you could add.

Endocrine system

The **endocrine system** consists of a network of **glands** that secrete chemical messengers – **hormones** – into the bloodstream (Figure 1.5.1). In the bloodstream, hormones are transported to receptors, which receive the information. The messages instruct organs to perform a variety of responses needed for homeostasis, sexual development and survival.

Hormones travel in the blood from a gland to a targeted receptor on a cell. Generally, hormones provide a slower communication system than electrical messages (nerve impulses), but they can have a longer effect. For example, hormones that are responsible for reproductive development cause a slow response compared with the instantaneous nervous response of shivering when cold.

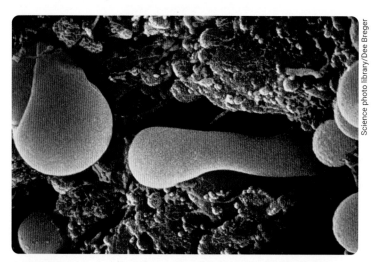

Science photo library/Dee Breger

▲ **FIGURE 1.5.1** A coloured scanning electron micrograph of the secretion of thyroid hormones from thyroid tissue. Hormones enter the capillary network (blue) and circulate throughout the bloodstream.

The main human endocrine glands

Are you still growing? During puberty, and for a few years afterwards, you may experience a series of growth spurts due to growth hormones released by the pituitary gland. Endocrine glands secrete hormones directly into the bloodstream. They regulate many human body functions, including growth. Figure 1.5.2 shows the location of the main endocrine glands: the hypothalamus, pituitary, thyroid, parathyroid, pancreas, thymus, gonads (testes in males and ovaries in females), pineal and adrenal glands.

Hormones

Hormones secreted by the hypothalamus, pituitary, thyroid, parathyroid, pancreas and adrenal glands are involved in negative feedback mechanisms. Hormones can be classified according to their chemical class, as either water-soluble proteins (long chains of amino acids), peptides (short chains of amino acids), amines (hormones made from a single amino acid) or fat-soluble steroids (made from the lipid cholesterol). The chemical class helps us predict their action when they arrive at a target cell.

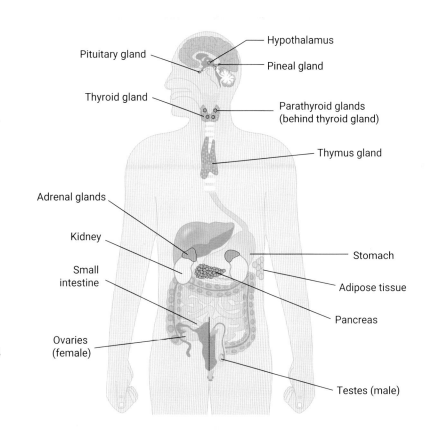

▲ **FIGURE 1.5.2**   The human endocrine glands (shown in yellow)

Some examples of hormones and their functions are listed in Table 1.5.1.

▽ **TABLE 1.5.1**   Some of the human endocrine glands, major hormones and their functions

Gland	Hormone example	Hormone class	Hormone function
Pituitary gland	Antidiuretic hormone (ADH)	Peptide	Controls water regulation by increasing reabsorption of water in kidneys
	Growth hormone	Peptide	Stimulates growth
Thyroid gland	Thyroxine	Protein	Controls metabolic rate
Pancreas	Insulin	Protein	Lowers blood glucose level
	Glucagon	Protein	Raises blood glucose level
Ovaries	Progesterone	Steroid	Regulates the menstrual cycle, prepares and supports the process of pregnancy
Testicles	Testosterone	Steroid	Develops and maintains male sexual characteristics
Pineal gland	Melatonin	Amine	Regulates the biological 'sleep–wake' rhythm

The main steps of how a hormone works are summarised in Figure 1.5.3.

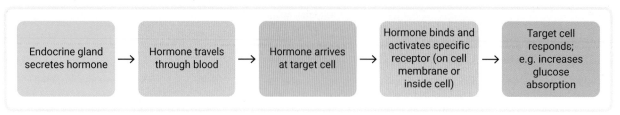

Endocrine gland secretes hormone → Hormone travels through blood → Hormone arrives at target cell → Hormone binds and activates specific receptor (on cell membrane or inside cell) → Target cell responds; e.g. increases glucose absorption

▲ **FIGURE 1.5.3** How a hormone works

Endocrine and nervous systems

The endocrine and nervous systems act individually and collaboratively to communicate and regulate body functions. The hypothalamus plays a major role in coordinating the endocrine and nervous systems. The hypothalamus receives sensory information from the nerves. In response, it uses a combination of endocrine and nervous messaging. The pituitary gland extends down from the hypothalamus, joined by axons. Hormones produced in the hypothalamus can be transported through the axons to the pituitary gland for storing.

Endocrine glands secrete hormones that travel through the blood to coordinate slower, longer-acting responses to stimuli. The nervous system uses fast, electrical nerve impulses along neurons. The two systems can work together to achieve homeostasis, development and reproduction.

1.5 LEARNING CHECK

1 **Describe** how a hormone works, step by step.

2 **Copy and complete** the table of glands and hormones by adding the missing information. Extend the table by adding three more glands and their hormone examples and functions.

Gland	Hormone example	Hormone function
Pituitary gland		Controls water regulation by increasing reabsorption of water in kidneys
	Growth hormone	Stimulates growth
	Insulin	
	Glucagon	

3 **Create** a table with one similarity and two differences between hormones and nerve impulses.

4 **Describe** the individual and collaborative roles of the endocrine system and nervous system.

9780170472838

Temperature regulation

BY THE END OF THIS MODULE, YOU WILL BE ABLE TO:
- ✓ describe several mechanisms that help control body temperature
- ✓ explain how and why body temperature is controlled
- ✓ draw a labelled negative feedback loop for temperature regulation.

GET THINKING

Scan this module for key science terminology to add to your growing mind map for this chapter.

Interactive resource
Drag and drop: Maintaining body temperature

Temperature variation

Humans are exposed to extreme variations in temperature. In 2022, the temperature in Australia ranged from –11.7°C to 50.7°C.

a

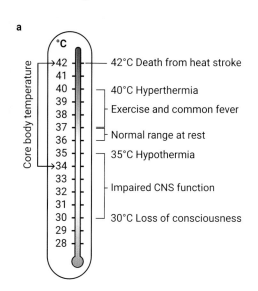

Core body temperature

- 42°C Death from heat stroke
- 40°C Hyperthermia
- Exercise and common fever
- Normal range at rest
- 35°C Hypothermia
- Impaired CNS function
- 30°C Loss of consciousness

Getty Images/Stringer/Brook Mitchell

c

AAP Images/STEVEN SAPHORE

▲ **FIGURE 1.6.1** (a) Even small temperature deviations from the body's core temperature can have significant impacts. Australia experiences extreme temperature variations from (b) hot to (c) cold.

Regardless of these changes in the external environment, our internal body temperature must stay close to 37°C. Small deviations away from this temperature can cause a range of medical issues (see Figure 1.6.1a). Just 5° above our normal 37°C can cause death from heat stroke. This means that humans have a relatively small tolerance range and a narrow temperature setting for optimal activity. How do we regulate our temperature to stay within such a small tolerance range?

Temperature regulation

The human body has several mechanisms for regulating internal body temperature. Behavioural mechanisms are simple actions we can take to change our temperature, such as moving into cold water to cool down or moving into a sunny spot and putting on a coat to warm up. Structural mechanisms are built-in physical features that help to regulate body temperature, such as a thick, **insulating** layer of hair on your head to keep your head warm. The mechanisms we will focus on for the rest of this chapter are physiological mechanisms. These are internal mechanisms that are automatically generated processes. They are controlled by our nervous and endocrine systems, which are coordinated by the hypothalamus.

Mechanisms that respond to a body temperature above 37°C include:

- sweating
- dilation of blood vessels
- a decrease in metabolic activity.

Mechanisms that respond to a body temperature below 37°C include:

- shivering
- constriction of blood vessels
- an increase in metabolic activity.

Sweating and shivering

An increase in temperature may cause sweat glands to open. When water and salt are released, the water draws heat from the body as it evaporates. As the water vapour moves into the surrounding air, it takes the heat energy away with it, cooling the body down. The effect is known as **evaporative cooling**. Figure 1.6.2 shows the location of sweat glands in the skin.

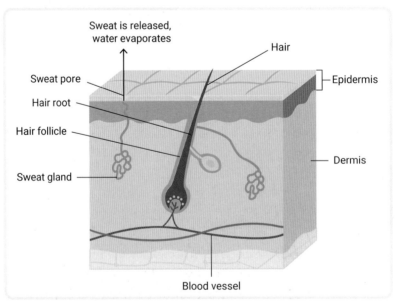

▲ **FIGURE 1.6.2**   Sweat glands in the skin release sweat. Water evaporates and has a cooling effect.

Shivering is a reflex action activated when we are very cold. Muscles rapidly contract and relax, generating heat.

insulating
reducing the transfer/loss of heat

evaporative cooling
the cooling effect that occurs when water evaporates

9780170472838

Negative feedback loop

A negative feedback loop can be drawn to represent the key steps in the process of maintaining an optimal internal body temperature. The hypothalamus has a set optimal temperature of 37°C. The hypothalamus responds to changes in the internal or external temperature by coordinating a response through the communication network of the nervous and endocrine systems. The systems send messages to effector muscles or glands to carry out a response that counteracts the stimulus. You can see examples of negative feedback loops in Figure 1.6.3.

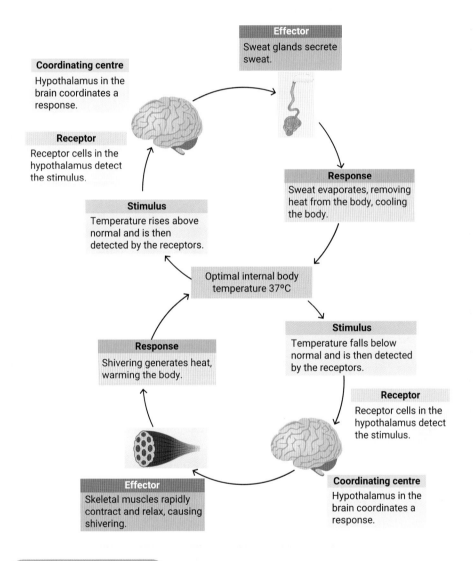

Coordinating centre
Hypothalamus in the brain coordinates a response.

Receptor
Receptor cells in the hypothalamus detect the stimulus.

Stimulus
Temperature rises above normal and is then detected by the receptors.

Effector
Sweat glands secrete sweat.

Response
Sweat evaporates, removing heat from the body, cooling the body.

Optimal internal body temperature 37°C

Stimulus
Temperature falls below normal and is then detected by the receptors.

Receptor
Receptor cells in the hypothalamus detect the stimulus.

Coordinating centre
Hypothalamus in the brain coordinates a response.

Response
Shivering generates heat, warming the body.

Effector
Skeletal muscles rapidly contract and relax, causing shivering.

◀ **FIGURE 1.6.3** A figure 8 model of two negative feedback loops for an increase or decrease in internal body temperature

---

1.6 LEARNING CHECK

1 **Define** 'shivering'.

2 **Explain** why internal body temperature in humans needs to be regulated.

3 **Create** a negative feedback loop for an increase in internal body temperature above 37°C, using dilation of blood vessels as the mechanism.

4 **Create** a negative feedback loop for a decrease in internal body temperature below 37°C, using constriction of blood vessels as the mechanism.

5 **Create** a crossword of 5–10 words for this module. Hand it to a partner to test it out.

1.7 Blood glucose regulation

BY THE END OF THIS MODULE, YOU WILL BE ABLE TO:

✓ describe several mechanisms that assist in the control of blood glucose levels

✓ explain how and why blood glucose is controlled

✓ draw a labelled negative feedback loop for blood glucose regulation.

Interactive resource
Label: Negative feedback loops

GET THINKING

You may know someone who has diabetes, a common condition that affects many Australians. What do you know about the condition? How do you think it relates to homeostasis? Write down your answers and then revisit them when you've completed the module.

Blood glucose variation

blood glucose
the amount of glucose in blood

Have you ever been 'hangry'? Hungry + angry = 'hangry'. Hunger is a stress response to low blood glucose. When we are hungry, our inhibition to other emotions, including anger, is lowered. **Blood glucose** is the amount of glucose in blood.

Glucose is required for cellular respiration, a metabolic reaction. Usually there are about 5 L of blood in an adult who weighs 75 kg. There are closer to 4 L of blood in an adolescent. The level of glucose in this volume of blood is regulated by the endocrine and nervous systems. The range of glucose in adult blood is 3.3–7 g. Glucose levels rise after every meal by 1–2 g. Glucose levels are lowest just after waking, which usually follows a night of fasting.

When glands are not correctly functioning to maintain levels within the tolerance range, glucose levels can rise to very high or fall to low. Persistently high blood glucose is known as hyperglycaemia. This may indicate low **insulin** production due to diabetes and can damage organs, such as the kidneys, nerves, retina and arteries. In contrast, when blood glucose is persistently low, symptoms such as sweating, shaking, drowsiness, seizures and unconsciousness can occur because of the inadequate supply of glucose to the cells and brain.

Getty Images/iStock/-101PHOTO-

▲ **FIGURE 1.7.1** 'Hangry' = hungry + angry

insulin
a hormone secreted by the pancreas; controls how much glucose is in the blood

glucagon
a hormone secreted by the pancreas; breaks down glycogen into glucose

Blood glucose regulation

The human body has several mechanisms to regulate blood glucose levels. Behavioural mechanisms include eating a healthy diet. We also have physiological mechanisms (internal processes) that are triggered by our nervous and endocrine systems.

Regulation of blood glucose is largely controlled by the endocrine system. The pancreas is a gland that detects blood glucose levels and secretes two types of hormones that can decrease or increase blood glucose: insulin and **glucagon**.

9780170472838

Stimulus: blood glucose too high

Varying amounts of insulin are secreted to keep blood glucose within a normal range. Insulin is secreted from beta cells in the pancreas. Insulin causes glucose to leave the bloodstream and enter the cells. It does this by binding to a target cell, which activates the transport of glucose across the cell membrane. Once inside cells, glucose can be used in cellular respiration. A rise in blood glucose is a stimulus that will be detected by the pancreas. In response, more insulin is secreted. Less insulin is secreted when blood glucose drops below normal, as shown in Figure 1.7.2.

Stimulus: blood glucose too low

An additional response to low blood glucose is the secretion of glucagon from alpha cells in the pancreas. Glucagon breaks down stored **glycogen** to glucose. Glucagon travels to the liver and attaches to liver cells. Liver cells convert glycogen into glucose molecules and release them into the bloodstream, increasing blood glucose levels (see Figure 1.7.2). A saying to help you remember the hormone that is released when glucose levels are low is: 'When the **gluc**ose is **gon**e, release **glucagon**'.

glycogen
a store of glucose in the liver and muscles

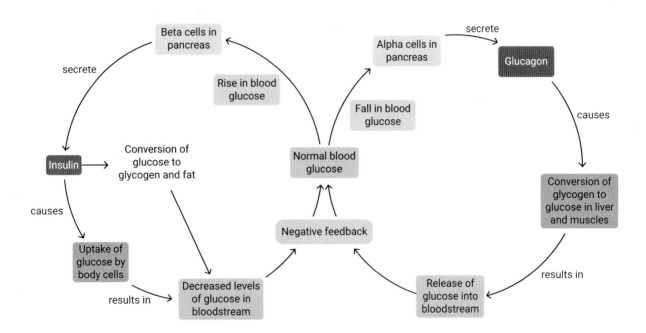

▲ **FIGURE 1.7.2** The negative feedback loops of glucose regulation

1 **State** the hormone released from:

  a alpha cells.   b beta cells.

2 **Define** 'blood glucose'.

3 **Draw** a negative feedback loop using insulin for an increase in blood glucose above normal.

4 **Analyse** Figure 1.7.2. Start at one point in the 'double' loop and write one sentence per stage to summarise the complete cycle of blood glucose negative feedback.

1.8 Use of technology to control blood glucose

BY THE END OF THIS MODULE, YOU WILL BE ABLE TO:
- ✓ describe diabetes monitoring and the use of data to make action plans
- ✓ investigate and evaluate how technologies and engineering enable advances in diabetes monitoring.

Video activity
What is diabetes?

Insulin – a vital hormone

Insulin is a vital hormone because it is the only hormone that can lower blood glucose levels. Insulin is produced by the beta cells in the islets of Langerhans in the pancreas. In a healthy person, insulin is secreted in response to rising blood glucose levels. The job of insulin is to attach to receptors on target cells to activate the transfer of glucose from the blood and into cells for use in cellular respiration.

What is diabetes?

About 1.8 million Australians have diabetes. People with diabetes cannot regulate their blood glucose levels because their bodies do not produce enough insulin. When they eat food containing glucose, the glucose stays in the blood instead of entering cells to make energy. This makes people with diabetes feel lethargic, dizzy and hungry. Long-term symptoms can be life-threatening.

There are three main types of diabetes: type 1, type 2 and gestational diabetes. Diabetes can be managed either through strict diets or by regularly injecting insulin (Figure 1.8.1).

Keeping blood glucose levels in a healthy range prevents short-term and long-term complications. Diabetes technology has developed over time, making the monitoring and treatment processes more accurate and less invasive.

Shutterstock.com/Orawan Pattarawimonchai

▲ **FIGURE 1.8.1** People with type 1 diabetes usually have to inject themselves with insulin several times a day to manage their blood glucose levels.

Blood glucose monitoring

A blood glucose monitor measures blood glucose levels. A small, needle-like device is used to create a puncture in the person's finger. Squeezing slightly ensures a small sample of blood appears. The blood is then transferred onto a test strip and, within 10 seconds, a result can be read and recorded in a journal. If the blood glucose value is too far away from normal, the person may need to inject insulin or have something to eat.

New technology and engineering now allow people to use the blood prick test three times a day or to monitor glucose continuously using a sensor inserted under the skin, in a process called flash glucose monitoring (Figure 1.8.2).

▲ **FIGURE 1.8.2 (a)** Finger-prick testing: a person punctures their finger to draw a drop of blood. **(b)** Flash glucose monitoring involves continuous monitoring with a sensor and compatible smartphone.

One benefit of flash glucose monitoring is that it collects continuous data, which can give more information about changes in blood glucose levels over time. This helps people living with diabetes to manage the disease better. In flash glucose monitoring, a sensor sits on the skin (on the back of the arm) and a small electrode is inserted just under the skin. A reader, such as an application on a smartphone, collects and provides a view of the data and trends. The reader can instantly share the data with a medical team. Alarms can be set to indicate when blood glucose levels are too high or low.

However, the flash glucose monitoring technology is not perfect. Finger pricks are still a more accurate monitoring method and are often preferred by doctors if glucose levels are changing too rapidly. Other potential drawbacks of flash glucose monitoring are:

- wearing a sensor may make it hard to do some sports such as swimming
- the electrode or sensor may irritate the skin
- it is more expensive to operate than the regular finger-prick test.

1.8 LEARNING CHECK

1 **Describe** the finger-prick glucose monitoring technique.
2 **Describe** the more technologically advanced flash glucose monitoring technique.
3 **Evaluate** the flash glucose monitoring technique, outlining the pros and cons.

Using a microscope to examine cells and collect data

SCIENCE SKILLS IN FOCUS

IN THIS MODULE, YOU WILL FOCUS ON LEARNING AND IMPROVING THESE SKILLS:

▶ selecting and using appropriate microscope equipment, including digital technologies, to record and communicate structures involved in homeostasis

▶ analysing and connecting a variety of data and information to identify and explain relationships.

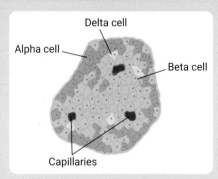

▲ FIGURE 1.9.1   A sketch of the islets of Langerhans

Video
Science skills in a minute: Using a microscope

Science skills resource
Science skills in practice: Using a microscope to examine cells

USING MICROSCOPES

▶ **Microscopes can magnify objects up to 400 times their actual size, which is why we use microscopes to see some cells. Using microscopes is an important science skill. When using a microscope, remember the following:**
- Carry the microscope with two hands.
- Click the lowest objective lens into place first. After focusing images on a low magnification, turn the revolving nose to click higher objective lenses into place.
- If your school owns a digital microscope, you can use it to capture and record images or videos on a computer. You can use your laptop if it is compatible.
- To calculate total magnification, multiply the objective lens magnification by the eyepiece lens magnification. Look for the values on the microscope parts.

▶ **Locating cells under a microscope is difficult. Use a combination of viewing specimens under the microscope and looking at photos of stained specimens.**

▶ **Draw and label a section of what you see, such as Figure 1.9.1. Figure 1.9.2 shows the labelled parts of the microscope.**

INVESTIGATION 1: MICROSCOPE INVESTIGATION OF THE PANCREAS

The pancreas is an endocrine gland that secretes hormones required for blood glucose regulation. The hormone-producing cells are located in sections called islets of Langerhans (discovered by anatomist Paul Langerhans in 1869).

AIM

To conduct a microscope investigation of the islets of Langerhans (the location of alpha and beta cells)

MATERIALS

☑ prepared and stained specimen of a pancreas section showing an islet of Langerhans
☑ access to the Internet
☑ light microscope
☑ digital microscope (optional)

METHOD

1   Set up your microscope by turning on the light and positioning the stage close to the objective lens.

2   Turn the revolving nosepiece until the lens with the lowest magnification clicks into place.

3   Place a prepared slide of an islet of Langerhans on the stage and use the stage clips to secure it in place.

4   Turn the coarse focus knob slowly until the image comes into focus. Without any further adjustments,

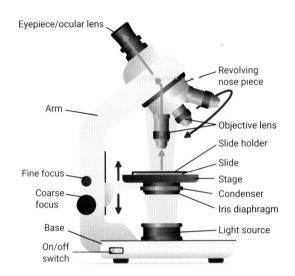

Eyepiece/ocular lens

Revolving nose piece

Arm

Objective lens

Slide holder

Slide

Fine focus

Stage

Coarse focus

Condenser

Iris diaphragm

Base

Light source

On/off switch

▲ **FIGURE 1.9.2** The parts of a microscope

turn the lens with the next highest magnification into position. Adjust the fine focus knob until the image in is focus.

5 Draw a sketch of what you see under low magnification, showing the distribution of islet cells. If the parts are stained, identify and label them on your diagram.

6 Focus a section of the specimen under high magnification. Draw a diagram of one islet, labelling what you can. Include a title and the total magnification next to each of your diagrams. You may see something like Figure 1.9.3.

7 Conduct an Internet search of at least 10 different microscope photos of islets of Langerhans. Draw one more sketch, one-third of a page in size, labelling the islet of Langerhans and alpha and beta cells.

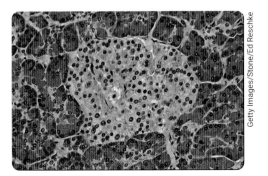

Getty Images/Stone/Ed Reschke

▲ **FIGURE 1.9.3** A microscope photo of pancreas cells. Islets of Langerhans are pale.

1 Describe the location of alpha and beta cells (where the hormones that regulate blood glucose are secreted from).

2 Describe what you could see under low power compared with what you could see under high power.

3 Evaluate the pros and cons of using a light microscope compared with using an electron microscope to look at these cells.

INVESTIGATION 2: INTERPRETING DATA AND COMMUNICATING RELATIONSHIPS BETWEEN VARIABLES

BACKGROUND

You will be studying the graphed results of a study conducted by specialist doctors. In the study, healthy participants fasted for 5 hours and then consumed glucose. Their blood glucose and insulin levels were monitored, and averages were recorded, over 5 hours.

AIM

To analyse and communicate trends in glucose and insulin data

METHOD

Analyse the two curves in Figure 1.9.4.

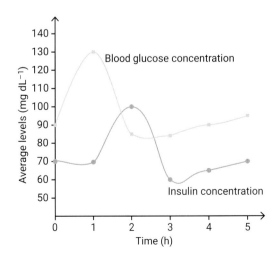

▲ **FIGURE 1.9.4** Mean (average) concentration of insulin and blood glucose 0–5 hours after glucose consumption

Chapter 1 | Homeostasis **29**

1 Describe the blood glucose concentration between 0 and 5 hours.

2 Describe the relationship between blood glucose and insulin levels.

3 Explain the cause of the insulin fluctuation between 2 and 4 hours.

INVESTIGATION 3: DISSECTING A BRAIN

AIM

To dissect a sheep's brain to observe the main structures visible

MATERIALS

- ☑ sheep's brain (slightly frozen if possible)
- ☑ dissecting equipment, including a scalpel and probe
- ☑ cutting board wrapped in newspaper
- ☑ gloves and apron
- ☑ safety glasses
- ☑ camera

Warning

Scalpels are sharp and cut easily. Carry the scalpel on a tray, always cut away from yourself, and keep the scalpel in the middle of the bench, not on the edge where it could fall. Wear covered shoes.

Biological material can contain disease-causing bacteria. Wear gloves and an apron and wash your hands thoroughly at the end of the experiment.

METHOD

1 Hold the brain carefully and locate the cerebrum, cerebellum, medulla oblongata and spinal cord.

2 Locate the lobes of the cerebrum, being aware that these are general areas, not discrete structures.

3 Take a photo of the brain. You will label the structures you located in step 1 later.

4 Gently feel the texture of the brain. Record your observations in a table.

5 Place the brain on your cutting board. Use the scalpel to carefully cut between the two hemispheres of the cerebrum. Try to locate the pituitary gland and hypothalamus. Take a photo of what you see.

6 Use the scalpel to carefully cut across the cerebrum. Identify the white matter and grey matter. Take a photo of what you see.

7 Repeat step 6 for the cerebellum.

8 Repeat step 6 for the spinal cord.

RESULTS

1 Draw a table to record your observations and photos.

2 Label the photos with the names of the parts that you identified. Use Figure 1.9.5 as guide.

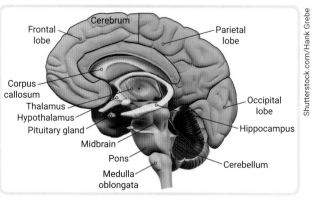

▲ FIGURE 1.9.5

EVALUATION

1 Suggest a reason for the soft texture of the brain.

2 Explain the reason for the difference between the white matter and the grey matter.

3 Reflect on what you thought the brain would look like. Does it look different? Name one characteristic that is different in structure and suggest why, in terms of its function.

1 Describe a stimulus–response model.

2 Copy and complete the following table by identifying or describing the functions of the parts of a negative feedback model.

Part	Function
Stimulus	
	A specialised cell that detects the stimulus. The receptor may be internal or external.
Coordinating (control) centre	
	A muscle or gland that receives a message from the coordinating (control) centre and carries out a response
Response	

3 Copy and complete the following table by identifying the three types of neurons according to their function.

Function	Type of neuron
Transmits nerve impulses from a sensory neuron to a motor neuron	
Detects changes and transmits nerve impulses towards the spinal cord or brain	
Transmits nerve impulses from the spinal cord towards a muscle or gland to carry out a response	

4 List the three parts of the brainstem.

5 Identify the following statements as true or false.

  a The cerebellum controls involuntary responses.

  b The brainstem controls involuntary responses.

  c The cerebrum controls muscle movements.

6 Draw a generalised model of a negative feedback loop.

7 Explain how the message in an electrical signal can get across a synapse.

8 Compare positive feedback and negative feedback and include an example of each.

9 State the internal body temperature tolerance range and optimal temperature for humans.

10 Describe the disease diabetes.

11 Read the following scenario and **identify** the type of feedback mechanism involved. **Justify** your answer by using the stimulus–response model.

After a run along the beach on a dry summer's day, Dave was feeling thirsty. The thirst centre in the hypothalamus part of his brain detected low water levels in his blood, so it sent a hormone to his kidneys to increase permeability (an increase in the ease of water flow), allowing water to return to his blood. This caused a decrease in the amount of water becoming part of his urine and moving into his bladder. When Dave went to the toilet, he found he only excreted a tiny amount of urine.

12 Explain how the structure of a sensory neuron allows for its function.

13 Compare the composition of white matter and grey matter in the brain.

14 Create a negative feedback loop for a decrease in internal body temperature below 37°C, using metabolic heat production as the mechanism. Use the negative feedback model in Figure 1.1.5 as a guide.

15 Explain how sweating works in regulating body temperature.

16 Explain the roles of insulin and glucagon in regulating blood glucose after a meal is consumed.

17 **Explain** why people who cannot produce sufficient insulin feel tired and lethargic.

ANALYSING

18 **Analyse** the following MRI image and locate and label eight parts of the brain.

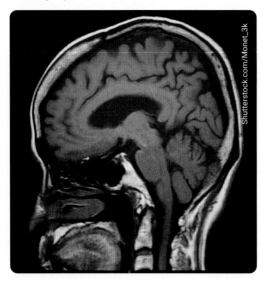

Shutterstock.com/MoneT_3k

19 The following diagram depicts two homeostasis processes.
   a Identify the two processes.
   b State which type of feedback loop is shown in each case.
   c Write out the steps shown in more detail, using the labels as subheadings.

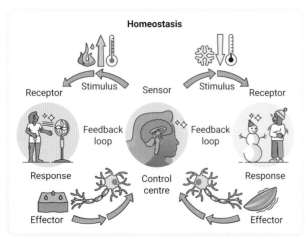

Homeostasis

Receptor — Stimulus — Sensor — Stimulus — Receptor
Feedback loop — Control centre — Feedback loop
Response — Effector — Response — Effector

EVALUATING

20 Look at the image of Robert Wadlow. He grew to be 2.52 m tall.

Alamy Stock Photo/History and Art Collection

   a How tall are you in comparison?
   b **State** the hormone that may have caused the excessive growth.
   c **Evaluate** whether there was too much or too little of the hormone secreted.
   d **Name** the gland the hormone was excreted from.

CREATING

21 **Produce** a short, engaging presentation for Year 7 students that explains what a negative feedback loop is, and how it is important in keeping us healthy. Include:
   • what the term 'homeostasis' means
   • information on the terms: stimulus, sensor, receptor, response and effector
   • an image of a negative feedback loop
   • examples of different negative feedback loops
   • reasons why negative feedback loops are important for keeping our body in a steady state.

22 **Create** a model in the form of a diagram or poster that demonstrates the flow of an electrical impulse along the three types of neurons between the PNS and the CNS.

BIG SCIENCE CHALLENGE PROJECT #1

1 Connect what you've learned

In this chapter, you have learned a lot about homeostasis and temperature regulation. Finish your mind map and reflect on how the concepts in each of the modules are related.

2 Check your thinking

Think back to when you last experienced a really hot day.

- Name one mechanism your body uses to regulate temperature.

- Can the body regulate its own temperature when the external temperature is very high?

- At which times of the day did you rely on an air conditioner to stay cool?

- Can you explain why our negative feedback systems are not effective in really hot weather?

- What are some smart ideas that can assist your negative feedback processes to keep you within human temperature tolerance range? Ideas could relate to cold drinks, clothing, houses and fitness.

3 Make an action plan

Make a list of your ideas for keeping cool and state how each idea helps a negative feedback process. For example, if you chose eating ice, you could explain that some internal heat will be used to melt the ice, cooling the body down.

4 Communicate

Convert your ideas into a labelled infographic to share with your class.

The immune system

2.1 Pathogens cause infectious disease (p. 36)

Infectious diseases are caused by pathogens entering a host. Bacteria, fungi and viruses can be pathogens.

2.2 Viruses (p. 39)

Viruses are non-living pathogens that replicate inside the cells of their hosts.

2.3 First line of defence against disease (p. 42)

Our bodies have ways of preventing pathogens from entering.

2.4 Second line of defence against disease (p. 44)

Swollen, red and painful injuries can be a good defence mechanism against infectious disease.

2.5 Third line of defence: adaptive immune response (p. 48)

When a pathogen enters our body, we adapt by coordinating an immune response, specific to the pathogen.

2.6 Artificial immunity (p. 52)

Vaccination stimulates our immune response without triggering the full effects of a disease.

2.7 FIRST NATIONS SCIENCE CONTEXTS: First Nations Peoples' use of medicinal plants (p. 56)

First Nations Australians have used plants for medicinal purposes for many thousands of years.

2.8 SCIENCE AS A HUMAN ENDEAVOUR: Using mosquito biology to fight the spread of disease (p. 60)

We use our understanding of the mosquito life cycle and anatomy to prevent the disease malaria.

2.9 SCIENCE INVESTIGATIONS: Petri dish safety (p. 62)

1 Efficacy of alcohol hand sanitiser
2 Modelling transmission of an infectious disease

9780170472838

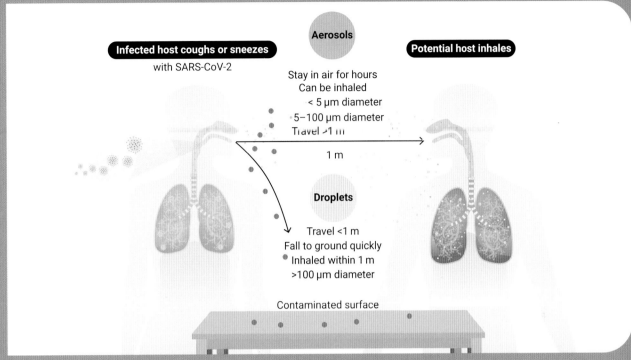

Aerosols

Infected host coughs or sneezes
with SARS-CoV-2

Potential host inhales

Stay in air for hours
Can be inhaled
< 5 µm diameter
5–100 µm diameter
Travel >1 m

1 m

Droplets

Travel <1 m
Fall to ground quickly
Inhaled within 1 m
>100 µm diameter

Contaminated surface

▲ **FIGURE 2.0.1** Airborne transmission of the SARS-CoV-2 virus, which causes the disease COVID-19 (1 µm (micrometre) = one-millionth of a metre)

Many infectious diseases are contagious. They are diseases that can spread (transmit) from person to person, such as COVID-19 and influenza. Infectious diseases spread easily between humans because there are many modes of transmission. Once you know how a disease transmits from an infected host to another person, you can slow or stop the spread.

▶ What are the two types of transmission of COVID-19 shown in the diagram? How else can COVID-19 be spread?

▶ What can we do to protect ourselves and people we come into contact with?

#2 SCIENCE CHALLENGE ACCEPTED!

At the end of this chapter, you can complete the Big Science Challenge Project #2. You can use the information you learn in this chapter to complete the project.

Assessments
- Prior knowledge quiz
- Chapter review questions
- End-of-chapter test
- Portfolio assessment task: Science research project

Videos
- Science skills in a minute: Petri dish safety (2.9)
- Videos activities: Immune defence (2.5); Small pox: the first vaccine (2.6); Facts about malaria (2.8)

Science skills resources
- Science skills in practice: Petri dish safety (2.9)
- Extra science investigation: Sterile technique (2.9)

Interactive resources
- Match: First line of defence (2.3); T cell or B cell? (2.5)
- Label: Virus replication (2.2); Phagocytosis (2.4)
- Drag and drop: Bacteria or fungi? (2.1)
- Quiz: Second line of defence (2.4); Herd immunity (2.6)

✷ Nelson MindTap

To access these resources and many more, visit:
cengage.com.au/nelsonmindtap

2.1 Pathogens cause infectious disease

Interactive resources
Drag and drop: Bacteria or fungi?

GET THINKING

Check this list of infectious diseases for ones you are familiar with. Have you suffered from any of them? Are there any that you haven't heard of?

- Influenza
- Conjunctivitis
- Whooping cough
- Ross River virus infection
- COVID-19
- Measles
- Chickenpox

Infectious disease

pathogen
an organism or virus that can cause an infectious disease

transmission
the transfer of a pathogen from one organism or reservoir to a new host

host
an organism infected with a pathogen

reservoir
a source of infection such as a habitat or an organism where a pathogen can replicate or survive for long periods

'Disease' is the term used to describe when an organism's body, or organs, cannot function normally. An infectious disease is transmissible and is caused by a **pathogen**. Any agent that causes an infectious disease in an organism is known as a pathogen. Disease-causing agents include micro-organisms, such as bacteria, fungi and protists, and non-living agents, such as viruses. We will look at viruses in more detail in Module 2.2. Not all micro-organisms are pathogens.

Transmission refers to the passing of a pathogen from an infected **host** or **reservoir** to a new host. A host is an organism infected with a pathogen. An infection occurs when a pathogen enters a host, establishes itself and replicates, causing symptoms. If a pathogen replicates but there are no symptoms, the infected host is described as asymptomatic but contagious. If transmission is direct from an infected host to a new host, the disease is contagious. A source of infection can be from another habitat or organism (a reservoir) where a pathogen can replicate or survive for long periods.

Types of pathogens

Three types of micro-organisms can cause infectious disease – bacteria, fungi and protists. They each have unique structures that help identify them. Scientists study a pathogen's structure and life cycle to make informed decisions about treatment and managing outbreaks.

Bacteria

Bacteria are unicellular, which means they consist of only one cell. An average bacterium is a microscopic 1–10 micrometres (μm) in length. Like all cells, bacteria have a membrane that surrounds the cytosol, and a cell wall. The cytosol is the semi-liquid substance inside all cells. Bacteria are prokaryotes. A prokaryote has no membrane-bound organelles such as mitochondria and a nucleus. This means the chromosome is found in the cytosol. Some bacteria possess a tail-like flagellum that helps them move.

Tuberculosis (TB) is an example of a disease caused by bacteria (Figure 2.1.1). In 2021, TB killed approximately 1.5 million people. TB is thought to be the world's oldest **pandemic**. Symptoms of TB include coughing (with mucus or blood), chest pains and fever. Infection occurs by inhaling airborne droplets containing the bacteria. The bacteria reproduce inside host cells, resulting in the symptoms. Active TB can cause serious illness and death. Reproduction involves a single cell dividing into two identical daughter cells.

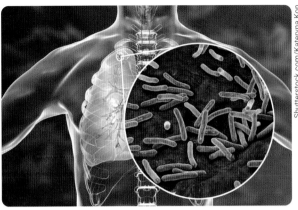

▲ **FIGURE 2.1.1** Tuberculosis is a disease caused by a bacterial pathogen.

pandemic
a disease that has spread rapidly across the world, having grown from an epidemic

Fungi

You may have seen the white fungus that grows on bread that looks like white, fuzzy cotton. The diverse fungal world includes large organisms, such as mushrooms, as well as minute forms that were only revealed with the invention of the microscope. Fungi have a eukaryotic cell structure with membrane-bound organelles, including mitochondria and a nucleus. They also have a cell wall made of a fibrous substance called chitin.

Moist, warm parts of a body provide ideal conditions for a fungal disease called tinea (athlete's foot) (Figure 2.1.2). Symptoms include an itchy, scaly rash, and peeling and inflamed skin. This contagious disease can be transmitted by direct contact with infected skin or contaminated surfaces such as the floor or shoes.

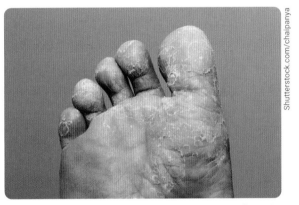

▲ **FIGURE 2.1.2** Tinea is a disease caused by a fungal pathogen. The warm, moist environment between the toes provides ideal conditions for reproduction.

Protists

Usually microscopic, protists can look very different from each other, and can be animal- or plant-like. They are eukaryotic organisms. Most protists are single celled and usually live in water.

The disease malaria is caused by the protist *Plasmodium*.

Methods of pathogen transmission

Pathogens infect a host by transferring from a source. A source could be an infected person or a reservoir. Respiratory (lung) diseases such as tuberculosis, influenza and COVID-19 can be transmitted when a person inhales contaminated airborne droplets. If another living thing transports and assists the entry of a pathogen into a host, from another host or reservoir, it is known as a **vector**. A common example of a vector animal is a mosquito. Some diseases can be transmitted by several means. For example, COVID-19 can be transmitted by direct contact (touch) or through the air via droplets or **aerosols**.

vector
an organism that transmits a pathogen from an animal or a plant to another animal or plant

aerosol
fine droplets of saliva or mucus containing pathogens

Some major modes of pathogen transmission are summarised in Figure 2.1.3.

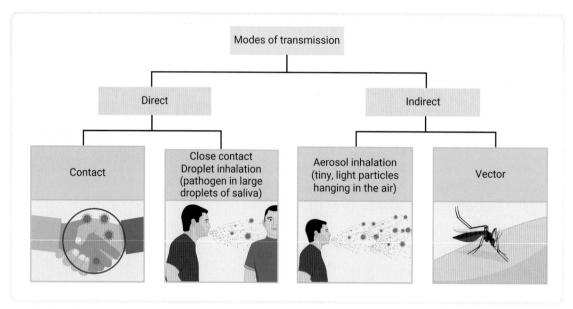

▲ FIGURE 2.1.3   Some major modes of transmission of pathogens

2.1   LEARNING CHECK

1 **Define** the terms 'pathogen' and 'micro-organism'.
2 **Describe** the disease tuberculosis (TB), including type of pathogen, symptoms and mode of transmission.
3 Study Figure 2.1.4, which shows a series of photos taken during an experiment on the effectiveness of different face masks in protecting against airborne pathogens.

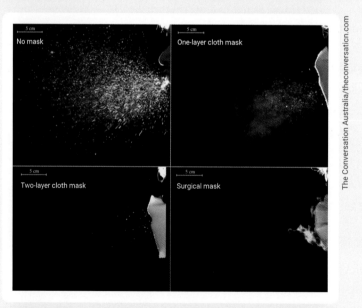

The Conversation Australia/theconversation.com

▲ FIGURE 2.1.4   An experiment testing the effectiveness of different types of masks when sneezing

State the dependent and independent variables. Write a statement that summarises the results. State one controlled variable. State a variable you think was not controlled that affected the validity.

9780170472838

GET THINKING

Viruses are very different types of pathogens from bacteria and fungi. Do you know how they are different? In 2 minutes, write down all the things you already know about viruses.

Interactive resource
Label: Virus replication

What are viruses?

Viruses are non-living pathogens. We can only see viruses if we use an electron microscope because they are so small, around 30–300 nanometres long (1 nanometre (nm) = one-billionth of a metre). The size and shape of viruses vary greatly, as shown in Figure 2.2.1.

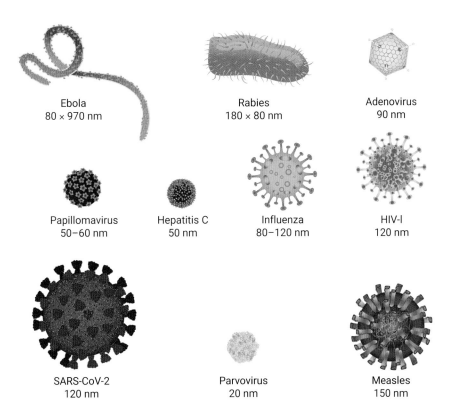

Ebola
80 × 970 nm

Rabies
180 × 80 nm

Adenovirus
90 nm

Papillomavirus
50–60 nm

Hepatitis C
50 nm

Influenza
80–120 nm

HIV-I
120 nm

SARS-CoV-2
120 nm

Parvovirus
20 nm

Measles
150 nm

▲ **FIGURE 2.2.1** Viruses are very diverse. These are some of the most common human viruses with their relative size.

There are thought to be more than 200 different types of viruses that can infect humans and cause disease.

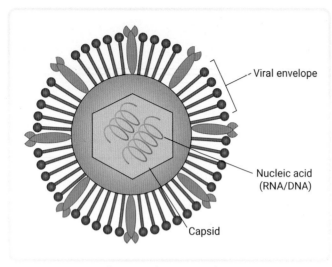

▲ **FIGURE 2.2.2**　The general structure of viruses

Labels in figure: Viral envelope; Nucleic acid (RNA/DNA); Capsid

Viruses consist of genetic material (nucleic acid), either **DNA** or **RNA**, covered by a protective layer called a protein **capsid**. The basic structure of a virus is shown in Figure 2.2.2. In some viruses, the capsid is protected by another layer called a **viral envelope**.

Viruses can only replicate (produce new viruses) inside a host cell. When a virus infects an organism, it recognises a receptor on the surface of a host cell and then attaches. It then enters through the cell membrane and inserts its DNA or RNA into the host cell. Once inside, the nucleic acid takes control and directs the host cell to make many copies of the virus parts. The parts are assembled into new viruses and released by the cell. This can cause the cell to die. The host usually experiences symptoms when a virus is replicating. Figure 2.2.3 represents the main steps of how a virus replicates inside the host cell.

DNA
deoxyribonucleic acid, the molecule that makes up the genetic material inside cells and in some viruses

RNA
ribonucleic acid, a molecule similar to DNA; the genetic material found in many viruses

capsid
the protein coating that protects the genetic material inside a virus

viral envelope
the outer layer of many viruses

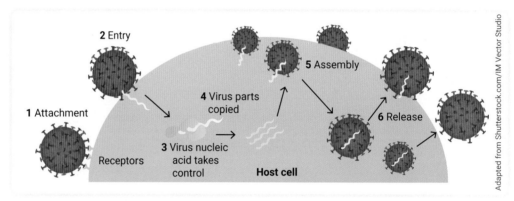

Labels in figure: 2 Entry; 5 Assembly; 4 Virus parts copied; 1 Attachment; 6 Release; 3 Virus nucleic acid takes control; Receptors; Host cell

Adapted from Shutterstock.com/IM Vector Studio

▲ **FIGURE 2.2.3**　How a virus replicates inside a host cell

COVID-19

COVID-19 is a viral disease that has spread globally. The World Health Organization (WHO) declared the disease a pandemic on 11 March 2020, three months after it was first reported. COVID-19 is caused by severe acute respiratory syndrome coronavirus 2 (SARS-CoV-2) (Figure 2.2.4).

▶ **FIGURE 2.2.4**　An electron microscope photo (micrograph) of SARS-CoV-2 virus particles

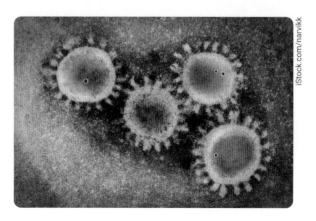

iStock.com/narvikk

The viral genome has had many variants since 2019. Variants arise when some of the RNA mutates. A mutation is a permanent change in the nucleic acid that codes for the production of virus parts. WHO names the variants after letters of the Greek alphabet, such as alpha, beta, gamma, delta and omicron.

The most common symptoms of COVID-19 are:

- fever
- tiredness
- cough
- loss of taste or smell
- headache
- shortness of breath
- sore throat
- muscle or joint pain.

Most infected people develop mild to moderate illness. Older people, or those with compromised immune systems, may develop severe disease symptoms, such as difficulty breathing and chest pain. It can take 5–14 days from when someone is infected to develop symptoms. This is called the incubation period. During this period, the infected person is contagious. This is the reason why, for much of the pandemic, infected people and their close contacts were often asked to quarantine (isolate themselves) or wear a mask.

Ross River virus

Ross River virus infection is a disease caused by a virus, delivered to humans by mosquitoes (vectors) when they feed on our blood. Symptoms include fever, chills, headaches, swollen joints and muscle pain. A rash may occur for 7–10 days (Figure 2.2.5).

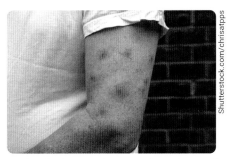

▲ FIGURE 2.2.5   Ross River virus infection is transmitted to humans by mosquitoes (vectors).

2.2 LEARNING CHECK

1 **Describe** how viruses replicate and spread.

2 Write two or three sentences to **explain** why viruses are classified as non-living.

3 Choose one virus and one bacterial pathogen to research.

   a **Draw** a diagram of each and **label** its parts.

   b Copy and complete the table to **compare** them.

	Bacteria example	Virus example
Size and shape		
Living or non-living		
How it reproduces		
How it is transmitted		

2.3 First line of defence against disease

BY THE END OF THIS MODULE, YOU WILL BE ABLE TO:
- ✓ list and label external defences of the body
- ✓ explain how the skin, stomach acid, cilia and secretions prevent infectious disease.

Interactive resource
Match: First line of defence

GET THINKING

Prepare a concept map, similar to the one below, to summarise the content in the modules you have studied so far. Add to it after you finish each module. You may wish to include diagrams within the map to help your understanding.

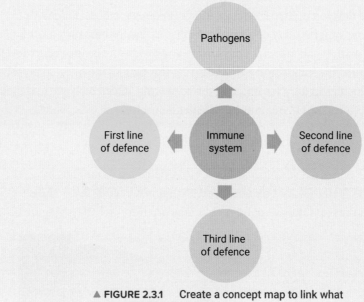

▲ **FIGURE 2.3.1** Create a concept map to link what you have learned so far.

The immune system's three lines of defence

The human immune system has three main lines of defence against infectious disease. The first and second lines of defence are non-specific, which means they defend against *any* foreign bodies or invading pathogens. We will look at the second line of defence in more detail in Module 2.4.

In Module 2.5, you will learn about a third line of defence – the adaptive and specific immune response. This is when highly specialised cells and antibodies defend the body against specific pathogens.

The first line of defence

The first line of defence involves prevention. Prevention is a fundamental principle of health care because it is easier for your body to prevent disease than it is to cure disease. The first line of defence consists of four main parts: the skin, stomach acid, **cilia** and secretions (which includes **lysozymes** found in tears and **mucus** found in nasal secretions). These help prevent the entry of pathogens and can be seen in Figure 2.3.2.

cilia
a hair-like extension of mucous membranes that pushes pathogens up and out of the body

lysozyme
an enzyme that breaks down pathogens

mucus
thick, sticky liquid in our airways that captures pathogens

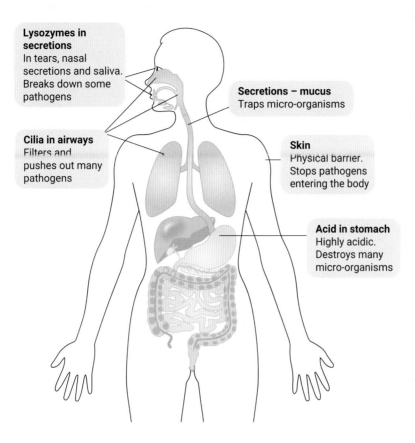

Lysozymes in secretions
In tears, nasal secretions and saliva. Breaks down some pathogens

Secretions – mucus
Traps micro-organisms

Cilia in airways
Filters and pushes out many pathogens

Skin
Physical barrier. Stops pathogens entering the body

Acid in stomach
Highly acidic. Destroys many micro-organisms

▲ FIGURE 2.3.2   Parts of the external, first line of defence of the immune system

External defences

Organs such as the lungs, intestines and stomach are considered external to the body. They are not protected by the adaptive immune response (third line of defence) because they cannot be reached by the specialised cells that are needed. However, as shown in Figure 2.3.2, parts of the digestive and respiratory systems play a part in the first line of defence.

2.3 LEARNING CHECK

1 **List** the four main parts of the first line of defence of the immune system.
2 **Draw** a human and label your drawing with the list you wrote for Question 1. Include the functions of those parts in your diagram.
3 There are four main reflexes of the human body that assist in the first line of defence: coughing, sneezing, vomiting and diarrhoea. Conduct some research on one to find out what function it serves in your body. **Write** 3−5 sentences about its function.

Second line of defence against disease

✓ describe and draw labelled diagrams of a phagocyte, a lysosome, phagocytosis, inflammation and fever

✓ list the symptoms of inflammation and explain how it protects the body.

Quiz
Second line of defence

Interactive resource
Label: Phagocytosis

phagocytosis
a process in which blood cells called phagocytes engulf and destroy a pathogen

inflammation
a response triggered by damaged cells in the body or a pathogen; characterised by redness, swelling, heat and pain

fever
a non-specific response to infection where the body temperature exceeds the normal 37°C

GET THINKING

Scan this module for any familiar terms. Have you had a wound swell, turn red and become painful? This module will help you understand why.

The second line of defence

If a potential pathogen gets past the external barriers and enters the body, the second line of defence commences. Just like the first line of defence, this second type of immune system involves a non-specific response. But this time, it is internal processes and cells that work to defend our body against invading pathogens. The processes of **phagocytosis**, **inflammation** and **fever** work together to defend the human body, as summarised in Figure 2.4.1.

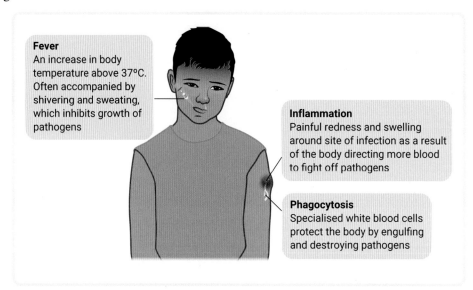

Fever
An increase in body temperature above 37°C. Often accompanied by shivering and sweating, which inhibits growth of pathogens

Inflammation
Painful redness and swelling around site of infection as a result of the body directing more blood to fight off pathogens

Phagocytosis
Specialised white blood cells protect the body by engulfing and destroying pathogens

▲ **FIGURE 2.4.1** Processes of the internal, second line of defence of the immune system

Phagocytes and phagocytosis

white blood cell
a specialised cell in the blood that defends the body against disease

phagocyte
a specialised white blood cell that can engulf and destroy pathogens

We have many different types of **white blood cells** as part of our immune system. Specialised white blood cells called **phagocytes** are like personal 'biological weapons' that can engulf and destroy pathogens in the process of phagocytosis.

There are different types of phagocytes in our blood, each with slightly different functions. However, they are all large white blood cells capable of identifying 'non-self' cells and changing their shape to surround an invading substance. As a phagocyte engulfs the cell, **lysosome** organelles digest (break down) the pathogen using enzymes. Figure 2.4.2 shows the main stages of phagocytosis.

lysosome
a cell organelle that contains digestive enzymes

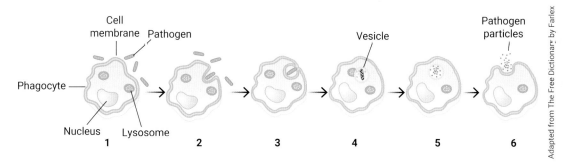

▲ **FIGURE 2.4.2** The process of phagocytosis

The process of phagocytosis consists of the following steps.

1 A phagocyte recognises a potential pathogen (e.g. bacteria).

2 The phagocyte changes its shape to surround the pathogen.

3 The cell membrane engulfs the pathogen.

4 The pathogen fuses with a lysosome, which injects a mix of enzymes to help digest the pathogen.

5 The pathogen breaks down into small particles.

6 The particles are released from the phagocyte.

Inflammation

Inflammation usually hurts, but it is a defence system that helps protect the body from pathogens. Inflammation is triggered by damaged cells in the body or by a pathogen that got past the barriers of the immune system's first line of defence (Figure 2.4.3). The inflammatory response sends for immune cells, such as phagocytes, to remove dead cells and begin repairing tissue.

The signs of inflammation are:

- redness
- swelling
- heat
- pain.

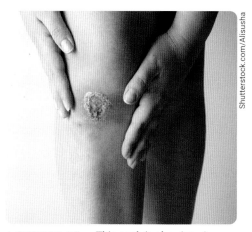

▲ **FIGURE 2.4.3** This scab is showing signs of infection.

Figure 2.4.4 shows the main stages of inflammation.

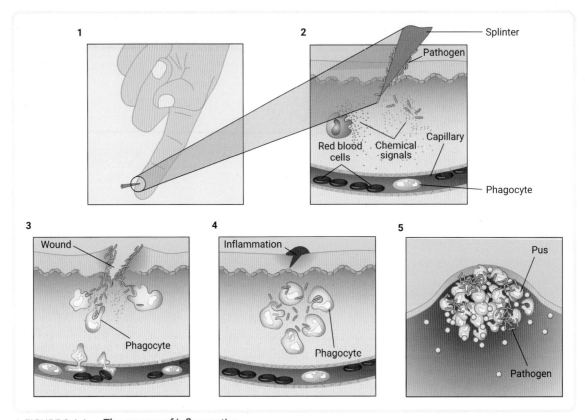

▲ **FIGURE 2.4.4**   The process of inflammation

The process of inflammation consists of the following steps.

1   The body suffers a wound, which is an entry point for pathogens.

2   Cells near the wound release chemical signals.

3   These cause blood vessels to dilate, and become more permeable, which increases blood flow near the wound. This results in redness, heat, swelling and pain.

4   Phagocytes arrive at the site of the wound to destroy the pathogens.

5   Pus (a mix of used phagocytes, dead cells and tissue fluids) may form but will eventually be removed by phagocytes.

Getty Images/E+/VioletaStoimenova

▲ **FIGURE 2.4.5**   A temperature above 39.0°C is considered a high fever.

Fever

Inflammation can also lead to fever, another non-specific response (Figure 2.4.5). Instead of a localised response, fever affects the whole body. Symptoms include feeling unwell, hot and sweaty and shivering. When internal temperature exceeds the normal 37°C, in response to an infection, the body is experiencing fever. A high fever (about 39°C) defends the body by:

• reducing the growth rate of the pathogen

• increasing the body's metabolic activity, which makes the immune cells respond faster.

9780170472838

Using a digital thermometer

Test the effectiveness of measuring body temperature by the 'under the arm' method with a digital thermometer. You will need to ask your teacher if you can use a thermometer. Conduct multiple trials on three people and calculate an average body temperature for each person. Disinfect the thermometer before using it on different people.

2.4 LEARNING CHECK

1 **Define** 'phagocyte'.
2 **Describe** the steps of phagocytosis.
3 **Draw** an annotated diagram of the steps of the inflammation response.
4 **Compare** (discuss the similarities and differences between) the first and second lines of defence in terms of:
   - the processes and parts of the body involved
   - whether they are specific or non-specific processes/parts of a disease.

   You can do this by using a Venn diagram, as shown below.

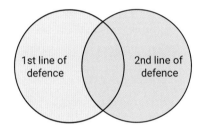

5 Write a paragraph to **explain** why processes in the second line of defence are described as 'non-specific'. Use examples to **justify** your reasoning.
6 Lysosomes are found in all animal cells. Figure 2.4.6 shows an animal cell with lysosomes.

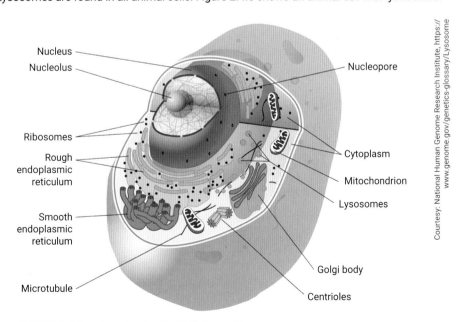

Courtesy: National Human Genome Research Institute, https://www.genome.gov/genetics-glossary/Lysosome

▲ **FIGURE 2.4.6** An animal cell with lysosomes

What is the function of the lysosomes? What type of human cell uses lysosomes in the second line of defence?

2.5 Third line of defence: adaptive immune response

BY THE END OF THIS MODULE, YOU WILL BE ABLE TO:

✓ classify and describe the function of the major white blood cells of the immune system

✓ identify where the immune cells are produced in the body

✓ define antigen, antibody, lymphocyte, T cell, B cell and memory cell

✓ describe the different parts of the adaptive immune system and how they work together.

Video activity
Immune defence

Interactive resource
Match: T cell or B cell?

antibody
a Y-shaped protein that binds to one type of antigen

antigen
a substance on a pathogen that stimulates the adaptive immune response

B cell
a type of white blood cell involved in the humoral (antibody)-mediated immune response

T cell
a type of white blood cell involved in the cell-mediated immune response

lymphocyte
a type of white blood cell involved in specific immune responses

GET THINKING

Have you had a cold recently? How long did it last? 7–10 days? That is approximately how long it took for your third line of defence to launch a specific response and destroy the pathogens.

The specific immune system

The specific immune system responds directly to a specific invading pathogen by producing an 'army' of immune cells and **antibodies**. It is also known as the adaptive immune system because the immune cells and antibodies need to adapt in structure and function to suit the specific **antigen** present on the pathogen.

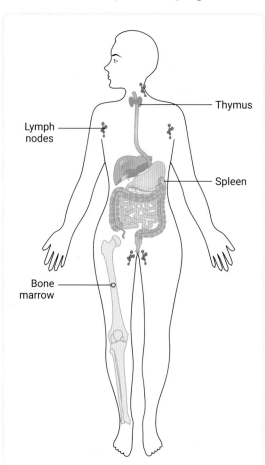

The names of our immune cells are B-lymphocytes (**B cells**) and T-lymphocytes (**T cells**). **Lymphocytes** are a type of white blood cell. B cells are produced and mature in the bone marrow (B for bone marrow). T cells are produced in the bone marrow too, but they mature in the thymus (so T for thymus). Check Figure 2.5.1 for the locations of the organs involved in the specific immune response.

The spleen contains phagocytes that engulf and break down old or dead cells from blood vessels. Lymph nodes are connected by lymph vessels that absorb and transport fluid that leaks out of blood vessels. The fluid inside lymph vessels is called lymph. Pathogens can travel out of blood vessels into lymph and collect in the lymph nodes, where a variety of immune cells (lymphocytes and phagocytes) can target them.

▲ **FIGURE 2.5.1** Organs involved in the specific (adaptive) immune system

Antigens and antigen-presenting cells

When pathogens get past the barriers and processes of the first and second lines of defence, they can be targeted by the specific immune system. How does the specific immune system know which cells to destroy? And how do the immune cells and antibodies know what specific structures to develop?

The answer to both questions is 'antigens'. All cells, including pathogens, have antigens (usually proteins) on their surface (Figure 2.5.2). Viruses and toxins contain antigens too. Antigens stimulate an immune response if identified by the immune system. There are specialised white blood cells that can identify 'non-self' antigens, which are antigens that do not belong in the body. The specialised white blood cells are called antigen-presenting

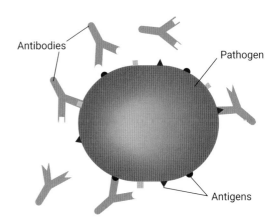

▲ FIGURE 2.5.2   Antibodies recognise and bind to specific antigens.

cells (APCs). APCs, such as phagocytes, identify, capture and present antigens to T cells and B cells. T and B cells are stimulated to adapt their structures, ready to defend against a specific pathogen. Some B cells produce antibodies, which are Y-shaped proteins that recognise and bind to a specific type of antigen on the surface of a pathogen.

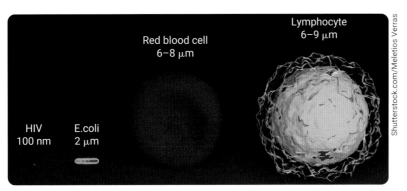

▲ FIGURE 2.5.3   A lymphocyte compared with a blood cell, a bacterial cell and a virus

A coordinated response

Have you watched a team of surgeons at work together in a movie? Each surgeon has a different role during the surgery. One surgeon applies anaesthetic, while another cuts and sews, but all work together to achieve the same purpose – to heal the patient. Our B cells and T cells are a team of specialised cells with different roles to achieve the same purpose – to destroy the pathogen and remember the antigen in case it invades again.

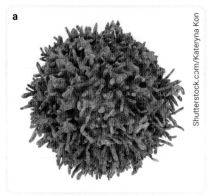

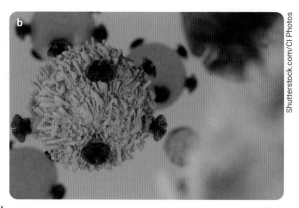

cell-mediated immunity
the specific immune response prompted by T cells on infected host cells

humoral immunity
the specific immune response prompted by B cells, and antibodies specific to the antigen, on pathogens found outside of host cells

▲ **FIGURE 2.5.4**   (a) A B cell and (b) a T cell

B and T cells mature into different types of specialised cells when providing a third line of defence (Figure 2.5.4). They are only activated when they recognise specific antigens on a pathogen. The main types of specialised T cells are helper T cells, cytotoxic (killer) T cells and memory T cells. The main types of B cells are plasma B cells (which produce antibodies) and memory B cells. T cells and B cells work together to provide a coordinated immune response (see Figure 2.5.5).

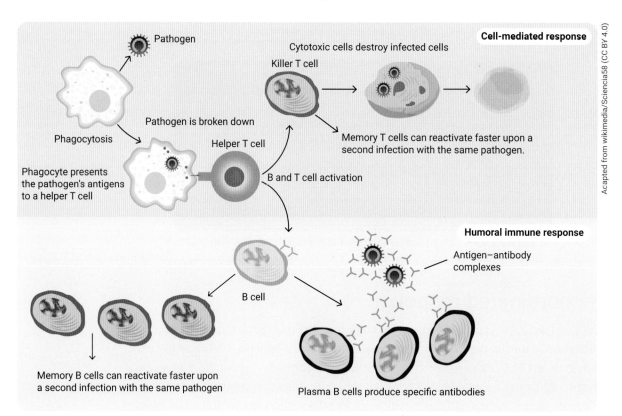

▲ **FIGURE 2.5.5**   The coordinated cell-mediated and humoral immune responses. This diagram shows the main functions of T and B cells.

The immune response generated by T cells is classified as **cell-mediated immunity**. The immune response generated by B cells is classified as **humoral immunity**. Table 2.5.1 summarises the main types and functions of B cells and T cells.

9780170472838

▼ TABLE 2.5.1 Types and functions of the main types of T cells and B cells

Type of cell	How are they activated?	Function and type of response
Helper T cell	Immature helper T cells mature when presented with an antigen	• Signal immature B cells to become plasma B cells • Cell-mediated immunity
Plasma B cell	Immature B cells mature into plasma B cells and multiply when signalled by helper T cells	• Make lots of specific antibodies that act on the antigen • Antibodies attach to the antigen on the surface of a pathogen and make it harder for the pathogens to invade a host cell • Humoral immunity
Cytotoxic (killer) T cell	Antibodies act like a marker for cytotoxic (killer) T cells to destroy the pathogen	• Release a toxic chemical when it encounters an infected host cell • The chemical destroys the damaged cell and the pathogen inside • Cell-mediated immunity
Memory T cell and B cell	Some of the specialised T and B cells form 'memories' of a specific pathogen	• Generate a faster and more intense immune response if a host is exposed to the same pathogen • B cells: humoral immunity • T cells: cell-mediated immunity

2.5 LEARNING CHECK

1 **Define**:
   a antigen.
   b antibody.
   c T cell.
   d B cell.
   e memory cell.

2 **Describe** the function of white blood cells.

3 **Identify** the correct cell type (B cell or T cell) for the following functions.
   a Humoral response
   b Produces antibodies
   c Marks pathogens ready for destruction

4 **Draw** your own immune system and label the main barriers, bone marrow, thymus, lymph nodes and spleen. Then **construct** a table to record the parts and their functions.

5 **Analyse** Figure 2.5.5. Write out the steps of the coordinated cell-mediated and humoral immune responses.

**BY THE END OF
THIS MODULE, YOU
WILL BE ABLE TO:**

✓ define and compare natural immunity and artificial immunity

✓ describe how vaccination stimulates the specific immune response

✓ describe the principles of herd immunity.

Video activity
Small pox: the first
vaccine

Quiz
Herd immunity

natural immunity
the protection obtained
from the production of
antibodies after exposure
to an infectious disease

memory cell
a mature B cell or
T cell that 'remembers'
information about a
specific antigen

vaccine
a substance that
stimulates a response in
the immune system and
produces antibodies

vaccination
administration of a
vaccine to provide
protection from a specific
disease

artificial immunity
the protection obtained
from the production of
antibodies after receiving
a vaccine or antibodies
found in medicine

GET THINKING

Imagine you have to convince someone that vaccines are important. Scan the information in this module and list as many facts as you can to support your argument.

Natural immunity

Natural immunity (specific immunity) can be achieved by the production of antibodies after the invasion of a pathogen. During the coordinated response, antibodies and **memory cells** are actively produced, specific to one antigen. This ensures the body is ready for a quicker and stronger immune response in the case of re-infection with the same pathogen. This usually provides long-term protection from reinfection.

Artificial immunity from antibodies or vaccination

To treat or prevent an infectious disease, scientists can inject antibodies from the blood of people who have recovered from the same disease. For example, a treatment for COVID-19 is the antibody drug sotrovimab. The drug works by binding to the protein spikes on the virus, which prevents it from replicating. This drug is used early in the course of an infection, to reduce the severity of the disease. This is a short-term treatment because the new host does not produce memory cells.

A longer-term solution is the administration of a **vaccine**.

When you get vaccinated, your immune system responds as if it had been naturally exposed to a pathogen. The vaccine stimulates the production of mature B cells and T cells, antibodies and memory cells specific to an antigen. A vaccine is a preparation of a weakened (attenuated) or inactivated (dead) micro-organism or virus or part of one, used to artificially induce immunity to a specific disease (Figure 2.6.1). **Vaccination** is usually administered by injection, but some vaccines are given by mouth as a liquid or tablet.

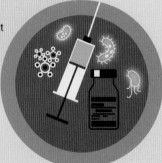

Inactivated vaccine
Consists of pathogen killed by heat
or chemicals, e.g. flu and polio

Toxoid vaccine
Consists of a treated toxin
(harmful substance) made by the
pathogen, e.g. tetanus

Attenuated (weakened) vaccine
Consists of a living but weakened
version of the pathogen,
e.g. chickenpox, measles
and mumps

**Subunit and nucleic acid
vaccines**
Consist of a section of genetic
material that codes for a protein,
e.g. COVID-19

Adapted from Bharat Biotech

▲ **FIGURE 2.6.1**   The four main types of vaccines

Artificial immunity is generally safer than natural immunity because the symptoms that follow the vaccine are usually not as severe as those of the infection of a pathogen.

One of the greatest vaccine success stories is the eradication (elimination) of smallpox. Smallpox is a highly infectious and deadly disease that killed an estimated 300 million people in the 20th century alone (Figure 2.6.2). It is the only human disease that has been eradicated.

Immunisation for smallpox started in 1796 when Dr Edward Jenner inoculated a boy with a mild cowpox, and then a few months later with the smallpox virus. The boy did not develop the disease. It was not until the 1970s that the World Health Organization coordinated a global vaccination program for smallpox, which eventually resulted in its eradication. The last recorded case was in 1977.

In Australia, the national immunisation program provides vaccines for 13 different diseases for your age group and younger. You may have received a dose (booster shot) in Year 7 for human papillomavirus, diphtheria, tetanus and whooping cough.

Boosters

The live attenuated (weakened) vaccines, such as those used for measles and mumps, can provide a lifetime of protection after one or two doses. Other vaccines, such as tetanus, need to be given in additional doses, or **boosters**, because, over time, antibodies and memory B and T cells decline.

▲ **FIGURE 2.6.2** Smallpox before it was eradicated. The boy on the right and the girl on the left had previously been vaccinated.

COVID-19 vaccines

booster
an additional dose of a vaccine, designed to provide a higher level of protection when antibodies have decreased

COVID-19 emerged in 2019. Its rapid spread and severe impact on people meant that urgent protection was needed. As soon as the genome was sequenced (genetic material decoded), it was shared globally and collaboratively, allowing for more than 250 vaccines to be developed.

Vaccines need to go through a series of trials before they can be approved by the Australian Government (Figure 2.6.3). Scientists:

- conduct laboratory research
- test for safety and efficacy on animals (such as ferrets or mice)
- conduct human clinical trials to ensure the vaccine is safe and effective.

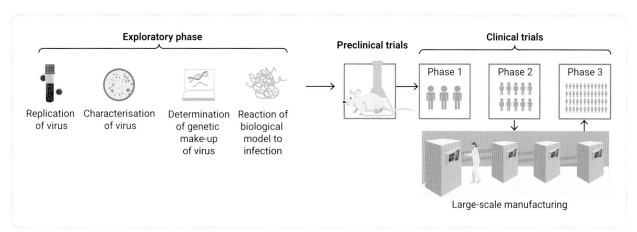

▲ **FIGURE 2.6.3** A vaccine needs to go through a series of steps before it can be approved.

Herd immunity

Herd immunity is a level of protection gained when a certain percentage of people in a community is immune to an infectious disease (Figure 2.6.4). This breaks the chain of transmission and stops the spread of disease. This means, if someone is infected in a community, but everyone else has been vaccinated and is immune, there will be no-one to spread the disease. Different diseases require different levels of herd immunity to stop their spread, depending on their transmissibility. For example, the vaccination threshold to stop the spread of measles is 95 per cent of the population, whereas polio has a threshold of 80 per cent. Natural immunity can lead to herd immunity; however, artificial immunisation by vaccination is a safer method.

Can we achieve herd immunity for COVID-19? It is a bit more complicated for COVID-19 than it is for many other infectious diseases because very contagious diseases like COVID need a higher threshold to achieve herd immunity. Additionally, herd immunity only works if a person can maintain their level of immunity. With COVID-19, we know people can still become infected even after a vaccine, and there is also evidence that people can get reinfected with different variants of COVID-19. This is because natural and artificial immunity wanes over time, or the vaccine is ineffective against new variants.

So why should we bother with COVID-19 vaccinations? A study by the US Centers for Disease Control and Prevention (CDC), conducted in 2021, showed that unvaccinated people are more than two times more likely to be reinfected with COVID-19 than those who are vaccinated. In early 2022, there was approximately a one in 1000 chance of being admitted to intensive care if you were double vaccinated, but a 15 in 1000 chance if you were unvaccinated. For most people, vaccination reduces the severity of COVID-19.

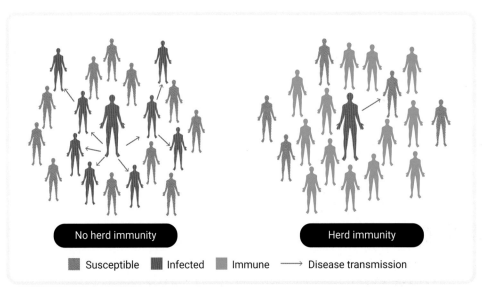

No herd immunity   Herd immunity

■ Susceptible ■ Infected ■ Immune → Disease transmission

▲ **FIGURE 2.6.4** Herd immunity reduces the transmission of disease.

1 **Define** 'vaccine'.

2 **Analyse** Figure 2.6.5. **Describe** the trend shown in this graph between vaccination and deaths and intensive care (ICU) admissions.

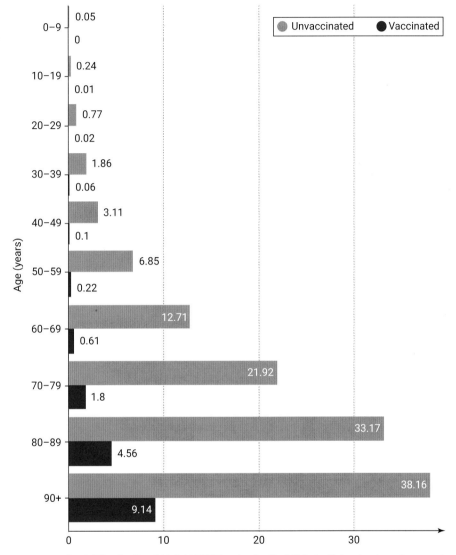

Copyright Guardian News & Media Ltd 2022, based on New South Wales health department surveillance reports

▲ **FIGURE 2.6.5** The proportion of COVID-19 cases with a severe outcome (death or ICU admission) by age and vaccination status from June 2021 to 1 January 2022

3 **Copy and complete** the following table to **compare** natural immunity and artificial immunity through vaccination.

Characteristic	Natural immunity	Artificial immunity (vaccination)
Specific or non-specific to a pathogen?		
Are antibodies made in response?		
Do memory cells form?		

2.7 First Nations Peoples' use of medicinal plants

IN THIS MODULE, YOU WILL:

✓ investigate native plant extracts used by First Nations Australians to treat infections and illness.

First Nations Australians' food and medicine

First Nations Australians have a wealth of scientific knowledge about native plants. For thousands of years, they have used plants for food and medicines and to develop tools and technologies (Figure 2.7.1). Medicinal use of plants requires knowledge about the right plants to use and how to prepare them to be effective.

▲ **FIGURE 2.7.1** Some First Nations Australians' foods: (a) lilly pilly, (b) Kakadu plum, (c) quandong, (d) Dianella, (e) seaberry, (f) native raspberry, (g) kangaroo apple, (h) native fig, (i) bunya nuts

Medicinal use of plants requires knowledge about the right plants to use and how to prepare them to be effective.

Native plants to treat coughs and cold

When you have a cold, you might use a medicated rub to soothe coughs and breathe more easily. Usually, the rub is covered with a warm cloth to help the medicines in the rub vaporise. First Nations Australians have been using medicated treatments for coughs and colds in this way for thousands of years. Prior to European colonisation, the Wiradjuri People (central New South Wales) built steam pits lined with eucalyptus leaves to treat coughs and colds. The eucalyptus vapour was released by heating the leaves over a fire with

▲ **FIGURE 2.7.2** Tea tree is used by the Yaegl People to treat the symptoms of coughs and colds.

possum rugs placed on top to create a steam pit. The Yaegl People (Coffs Harbour region, New South Wales) heat the leaves of tea tree (*Melaleuca alternifolia*) to relieve the symptoms of coughs and colds (Figure 2.7.2). Today, many common products that are used to treat coughs and colds contain eucalyptus or melaleuca oil.

Native plants to treat skin infections

There are many examples of plant extracts being used to treat or prevent skin infections. The Bundjalung Peoples (northern New South Wales/south-east Queensland) have long crushed tea tree leaves and applied them to skin wounds and infections, traditionally held in place with layers of paper bark (Figure 2.7.3).

▲ **FIGURE 2.7.3** Crushed tea tree leaves were traditionally held in place by paper bark to treat skin infections.

When European colonists first travelled to Bunjalung Nation, they were affected by skin infections caused by high rainfall, injuries and insect bites. Supplies of Western treatments were scarce. European colonists observed the Bunjalung Peoples' use of tea tree and adopted these methods to control infections. In the 1920s, scientist Dr A.R. Penfold investigated the antiseptic properties of tea tree. His study reaffirmed the knowledge of the Bunjalung Peoples, that the leaves contained powerful antiseptics that were more than 10 times stronger than other disinfectants of the time. The active components of tea tree have antimicrobial, antifungal, antiviral and anti-inflammatory properties. Tea tree grows mostly along the coastal areas of south-east Queensland and northern New South Wales.

First Nations Peoples in other parts of Australia use different plants to treat infections. The Peoples of the Torres Strait Islands have traditionally used coconut plant extracts to treat wound infections. Traditional Owners of the Northern Territory continue to use traditional medicine such as emu bush leaves (*Eremophila* species) to treat sores and cuts.

These examples give a glimpse of First Nations Australians' medicinal knowledge and use of native plants that continue today.

☆ ACTIVITY

Antimicrobial effect of products containing native plant extracts used by First Nations Australians

Follow your teacher's instructions. Be aware that some handwashes contain ingredients that can cause allergic or sensitivity reactions. Do not use these products if you have skin allergies or skin sensitivities. If you notice a reaction during this investigation, immediately wash your hands and seek medical advice.

Make sure your agar plates are securely taped once they have been swabbed.

You need
- ☑ 2 nutrient agar plates
- ☑ handwash containing tea tree or eucalyptus extract
- ☑ sterile swabs
- ☑ sterile water
- ☑ tape
- ☑ incubator

Warning
- Wear appropriate personal protective equipment (PPE).
- Do not open the Petri dishes after sealing them.
- Wash your hands thoroughly at the end of the investigation.

What to do

1 Dip a sterile swab into sterile water to moisten it.

2 Run the swab from your wrist at the base of your thumb up and over each of your fingers, down to the base of your wrist on the other side of your hand.

3 Roll the swab gently over the surface of one of the agar plates to cover the entire surface.

4 Tape the agar plate to seal it.

5 Use a handwash containing tea tree or eucalyptus extract to clean your hands. Make sure you rub the handwash between your fingers (Figure 2.7.4).

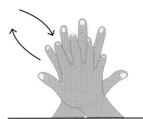 

Rub the palm of one hand over the back of the other hand with interlaced fingers. Repeat with your other hand.

Rub hands palm to palm with interlaced fingers.

▲ **FIGURE 2.7.4** How to wash your hands with handwash

6 Rinse off the handwash and wait for 5 minutes for your hands to air dry.

7 Repeat the hand swab in the same way as before, using a clean swab and rolling it onto the second agar plate.

8 Seal the plates with tape.

9 Incubate the plates at 37°C for 24–48 hours.

10 Count the number of colonies on the agar plates before and after treatment with the handwash.

11 Record your results in a table.

12 Calculate the percentage difference in the number of bacteria:

$$\text{Difference in number of microbial colonies (\%)} = \frac{\text{number of microbial colonies before handwash} - \text{number of microbial colonies after handwash}}{100}$$

What do you think?

1 Analyse the results and draw a conclusion about the effect of the handwash on skin microorganisms.

2 How do your results compare with those of other students in the class? Was one handwash more effective than the other?

3 Was this a fair test? Why or why not? What did you do to try and make this a fair test?

4 Propose another investigation in which the plant extract could be tested more specifically for its antimicrobial activity.

2.8 Using mosquito biology to fight the spread of disease

BY THE END OF THIS MODULE, YOU WILL BE ABLE TO:

✓ investigate how the values and needs of society influence the focus of scientific research.

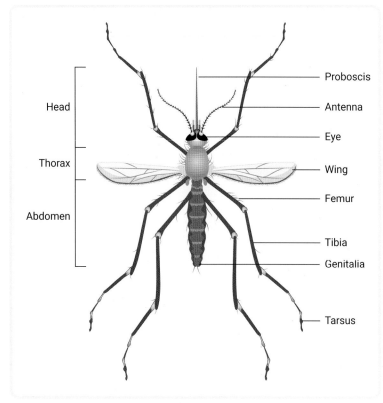

▲ **FIGURE 2.8.1**   The body parts of a mosquito

Labels: Head, Thorax, Abdomen, Proboscis, Antenna, Eye, Wing, Femur, Tibia, Genitalia, Tarsus

Video activity
Facts about malaria

Malaria

There were 241 million cases of malaria in 2020, most of them occurring in Africa. On average, more than 400 000 people die from malaria every year.

Current strategies for reducing the spread of mosquito-borne diseases are not effective enough to significantly reduce the number of deaths associated with these diseases. As a result, millions of dollars are spent every year on research into the prevention and treatment of malaria.

How is malaria transmitted?

Malaria is an example of a disease that is transmitted to humans by a vector, the *Anopheles* mosquito. Malaria only occurs in parts of the world where these species of mosquito are found. As for all species of mosquito, it is only the female mosquito that feeds on humans, because they need blood to feed their growing eggs.

The mosquito's mouth part, known as a proboscis (Figure 2.8.1), is a set of six needle-like darts that pierce the skin's blood vessels. Once pierced, the mosquito sucks blood, leaving behind saliva that can contain pathogens. The pathogen that causes malaria is a protist known as *Plasmodium*. The pathogens can then enter the bloodstream of the human and cause disease. People infected with malaria initially suffer fever and headache; if left untreated, severe illness can develop, followed by death.

Methods of preventing malaria

Do you use any of these strategies to prevent mosquito bites?

- Avoid outdoor activities at dawn and dusk.
- Use insect repellent.
- Remove stagnant water (mosquitoes breed in still water).
- Wear long clothing at dawn and dusk.
- Use bed nets and insecticide-treated bed nets.

Why do we use these strategies? Our knowledge of the mosquito's anatomy and life cycle helped us determine which strategies work best to prevent the spread of disease.

For example, mosquitoes lay their larvae in still water (Figure 2.8.2). This is why mosquito populations increase dramatically when it rains. If still water is treated or removed, the larvae cannot survive, resulting in fewer mosquitoes and reduced transmission of diseases such as malaria.

Is mosquito genetics the answer?

Scientists have decoded the genetic material of the whole mosquito genome. This opens up possibilities for many areas of research. For example, scientists are studying the genes involved in insecticide resistance. They have even produced a genetically modified mosquito by inserting two new genes. One of the genes is a lethal gene that gets passed to offspring and results in their death before they reach adulthood. The other gene is a fluorescent marker that allows scientists to keep track of the mosquitoes whose genomes contain the lethal gene. Many trials that involve releasing sterile males into populations in mosquito-infested areas of the world are currently under way and being monitored.

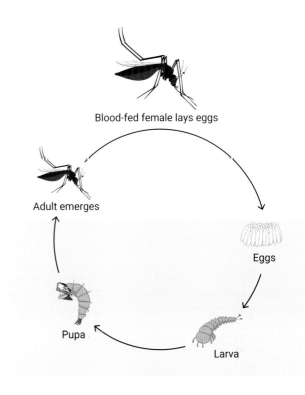

Blood-fed female lays eggs

Adult emerges

Eggs

Pupa

Larva

▲ **FIGURE 2.8.2** The stages in a mosquito life cycle

2.8 LEARNING CHECK

1 **State** the four stages of the life cycle of a mosquito.
2 **Explain** why female mosquitoes need to drink human blood.
3 The Bill and Melinda Gates Foundation has spent $1.5 billion on malaria research and control. **Explain** how this is an example of society's values and needs influencing scientific research.
4 Should we spend more or less money on malaria control research? **Justify** your answer.
5 **Describe** the structure and function of the mouthpiece of a mosquito and relate the structure to its ability to be a vector of an infectious disease.

2.9 Petri dish safety

SCIENCE SKILLS IN FOCUS

IN THIS MODULE, YOU WILL FOCUS ON LEARNING AND IMPROVING THESE SKILLS:

▶ assessing the risk involved in an investigation
▶ safely planning and conducting a Petri dish investigation when growing micro-organisms.

PETRI DISH SAFETY

▶ **Micro-organisms can cause disease. Most micro-organisms collected in the environment are not harmful. However, once they multiply, they may become a hazard.**
▶ **When you assess the safety requirements of a Petri dish investigation remember the following.**

1 It is only safe to grow micro-organisms in a Petri dish if the dish is sealed after you swab.
2 Dispose of your swab (cotton bud) straight after use in a bin.
3 Wash your hands straight after sealing the Petri dish.
4 When incubation is complete, do not open the lid of the Petri dish.
5 Observe the results with the lid still sealed.

Video
Science skills in a minute: Petri dish safety

Science skills resource
Science skills in practice: Petri dish safety

Extra science investigation
Sterile technique

INVESTIGATION 1: EFFICACY OF ALCOHOL HAND SANITISER

> AIM

To investigate how effective alcohol hand sanitiser is at killing micro-organisms

> BACKGROUND

People are encouraged to use more hand sanitiser to reduce the spread of COVID-19. Theoretically, hand sanitiser containing alcohol stops the spread of disease by breaking down the outer lipid layer of the virus.

> PREDICTION

Fewer micro-organisms will grow on a Petri dish if a swab is taken from a surface cleaned with hand sanitiser.

> MATERIALS

☑ 3 Petri dishes
☑ marking pen
☑ sticky tape
☑ cotton buds
☑ hand sanitiser containing 60–80% alcohol
☑ hand sanitiser or hand wash that does not contain alcohol
☑ safety glasses

> METHOD

Warning
• Wear appropriate personal protective equipment (PPE).
• Do not open the Petri dishes after sealing them.
• Wash your hands thoroughly at the end of the investigation.

1 Put on your safety glasses. Collect three Petri dishes containing pre-prepared nutrient agar. One will act as a control. Label each in small writing so that the writing does not obscure your observations.

Label Petri dish 1 'Control'. Label Petri dish 2 'Surface cleaned with alcohol hand sanitiser'. Label Petri dish 3 'Surface cleaned with non-alcohol hand sanitiser'.

Leave the cover on the plates at all times unless adding swabs. Only open the lid of the dish wide enough so that you can insert the swab or forceps.

2 Find a heavily used surface such as a handrail, drink fountain or keyboard. Use a clean cotton bud to get a swab of micro-organisms from this location. Do this by rubbing the cotton bud several times over one-third of the surface section you have chosen. Streak the first Petri dish (control) with the sample in a zig-zag motion, as shown in Figure 2.9.1.

3 Use the alcohol-based hand sanitiser to clean one-third of the chosen surface. Use another cotton bud to collect micro-organisms from the same location and then streak the second Petri dish with your sample in a zig-zag motion. Try to swab the same amount into each Petri dish.

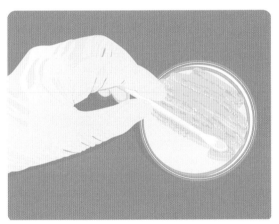

▲ FIGURE 2.9.1   Swabbing on agar in a Petri dish

4 Use a non-alcohol-based hand sanitiser to clean the final third of the surface. Use another cotton bud to collect micro-organisms from this location and then streak the third Petri dish with your sample in a zig-zag motion.

5 Draw a table to record the observed results.

6 Seal each dish with clear sticky tape and hand them to your teacher for incubation. Incubate at (37°C) for 2–3 days. To prevent water condensation accumulating and affecting your investigation, place the Petri dishes upside down where your teacher instructs you to.

7 Pack up and then wash your hands thoroughly.

RESULTS

Do not open the Petri dishes. Sketch or describe the results observed in each of the three Petri dishes.

Bacterial colonies are slightly circular, cream to yellowish in colour and shiny on the surface. Fungal colonies are more irregular and look like threads of cotton wool.

EVALUATION

1 **Describe** the purpose of the control.

2 Why was a non-alcohol hand sanitiser useful in this investigation?

3 **Record** the results of three other groups.

4 Write a conclusion for the investigation. Include a statement about whether the results supported your hypothesis.

5 **Evaluate** your safety precautions. Did you follow the steps outlined in Science skills in focus section? What could you have improved?

INVESTIGATION 2: MODELLING TRANSMISSION OF AN INFECTIOUS DISEASE

AIM

To model the transmission of an infectious disease

MATERIALS

- ☑ class set of cups containing 50 mL water
- ☑ cup containing 50 mL of dilute (0.1 M) sodium hydroxide solution
- ☑ class set of plastic droppers
- ☑ safety glasses

METHOD

Note: Your teacher will keep the identity of the cup with sodium hydroxide in it a secret, but also keep track of who collects it.

Warning
- Wear appropriate personal protective equipment (PPE).
- Sodium hydroxide and phenolphthalein can irritate eyes and skin. Wash with water if contact is made.

1 Wear safety glasses. Collect a clear plastic cup with 50 mL of liquid already poured in. All cups should contain the same volume of liquid.

2 One cup will contain sodium hydroxide (NaOH) solution. It is a clear colourless liquid that looks like water. It also has a high pH (very basic substance). The student who collects this cup will be 'patient zero'. All the other cups will contain water.

3 Collect a dropper.

4 When your teacher asks you to, walk slowly around the classroom until the teacher says 'stop'. Use the dropper to add 10 mL of your liquid into the cup of another student. Repeat this two more times until you have added to the cups of three other students. By doing this, you are modelling the spread of infection through body fluids such as droplets of mucus or saliva, which can contain pathogens.

5 Record who you shared your liquid with.

6 A volunteer student can use a dropper bottle with phenolphthalein indicator solution to test who has been infected. Phenolphthalein is an acid–base indicator. If the colourless liquid turns pink, the person is 'infected'.

▲ **FIGURE 2.9.2** Phenolphthalein indicator turns pink in basic solutions.

7 A table like the one below can be drawn on the board for all class members to contribute to. Add all names of students (givers) in the class in the first column, and then, highlight the 'infected' names to process the data.

Name of giver (potential source of infection)	Subject 1	Subject 2	Subject 3

8 Discuss and determine who could be patient zero (the original source of the 'infection').

1 **List** three different types of pathogens.

2 **Describe** how scientists classify a disease as infectious.

3 To which line of defence does inflammation belong?

4 **Identify** the type of biological molecule, and state the function of, a lysozyme found in tears.

5 **Copy and complete** the following table to identify the mode of transmission of each disease as direct contact, airborne or vector.

Disease	Information	Mode of transmission
Tuberculosis	Airborne droplets can be inhaled within 1 m.	
Malaria	An *Anopheles* mosquito has a blood feed and simultaneously injects the pathogen into the blood of a host.	
Smallpox		

6 **Explain** what antigens are and where they are found on pathogens.

7 **Explain** why a booster vaccine is required for some vaccination programs.

8 **Copy and complete** the following table to classify the different types of non-specific immune defence systems.

Immune system		System parts	Line of defence
Non-specific	External	1 Skin (secretions/barrier)	First
		2	
		3	
		4	
		5	
	Internal	1 Inflammation	Second
		2	
		3	

9 **Use** the term 'antimicrobial properties' to explain why certain First Nations Australian Peoples traditionally use tea tree leaves to treat wounds.

10 When humans suffer a skin wound, they may experience inflammation before healing, as shown in the diagram below.

List the signs of inflammation at the site of a wound.

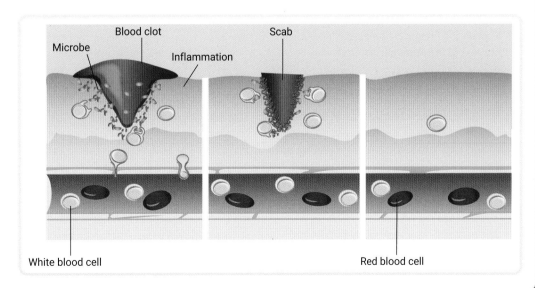

11 Mucous membranes occur in several parts of the body. The following image shows one of these areas, the lungs.

State the function of mucus in the human body.

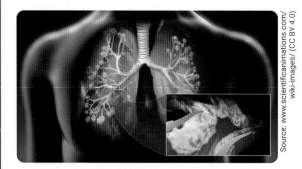

Source: www.scientificanimations.com/wiki-images/ (CC BY 4.0)

12 **Compare** natural immunity with artificial immunity by completing the table.

	Natural	Artificial
Is exposure to antigen deliberate or non-deliberate?		
Do antibodies develop?		
Does long-term immunity result?		

13 **Draw** a diagram showing the steps of phagocytosis.

Label your diagram with short statements to explain what happens at each stage.

ANALYSING

14 **Explain** the meaning of the term 'non-specific' in the context of the immune system.

15 **Explain** the impact of a break, or tear, in the skin on the body's defence against disease.

16 **Compare** the specific and non-specific immune systems. Include the major white blood cells involved.

17 A study was conducted at the University of New South Wales on how often medical students touch their face. The study found that:

- the 26 students touched their face an average of 23 times per hour
- just under half of all touches (1024 touches out of a total of 2345) were to areas that contain mucous membranes (eyes, nose or mouth).

a **Explain** why this behaviour might increase the spread of disease.

b **Predict** a message you think should be given to society about hand hygiene.

18 The image below represents herd immunity to a virus. The red figure represents an infected person. The grey figures represent unvaccinated people or those vulnerable to infection. The green figures represent people with either natural or artificial immunity. Imagine that each infected person is capable of infecting another two people. **Explain** what would happen if:

a the scenario remains the same

b the number of people with immunity was halved

c everyone had immunity.

EVALUATING

19 **Evaluate** the usefulness of fever in the immune response.

20 In 2020, scientists conducted experiments to determine how long SARS-CoV-2 (COVID-19) could remain infectious on a surface. One study found sunlight could 'kill' (denature) the virus within 14 minutes. By mid-2021, scientists had stopped studying contaminated surfaces and had turned to research and experiments on the effectiveness of wearing masks to slow the spread of the virus.

Explain why scientists may have changed their research focus. **Justify** your reasoning.

CREATING

21 **Create** a table summarising the structures of a bacterium, fungus and virus. Use three features for each type of pathogen.

22 **Design** your own poster summarising the cell-mediated and humoral immune systems.

BIG SCIENCE CHALLENGE PROJECT #2

At different times during the COVID pandemic, the number of hospitalisations has increased significantly. Sometimes there haven't been enough health workers to care for the high number of patients. How can we avoid overwhelming the hospital system?

1 Connect what you've learned

In this chapter, you have learned about pathogens that cause infectious disease, non-specific immune responses and the immune system. Finish the concept map you started in Module 2.3 and reflect on how the information in each of the modules is related.

2 Check your thinking

a Can you think of a disease caused by a:
- bacterium?
- fungus?
- virus?

b How do they enter a host and what are the symptoms?

c Is there a vaccine for each of them?

d Describe the modes of transmission for COVID-19. Use terms such as 'direct contact', 'indirect contact' (contaminated surfaces), 'close contact' and 'aerosol transmission'.

3 Make an action plan

Make a comprehensive list of actions (at least five) you can do to slow the spread of COVID-19 and any other airborne disease. For each action, provide an instruction on how to do the action effectively.

4 Communicate

Convert your ideas into a labelled infographic to share with your class, like the one on the right from SA Health on stopping the spread of gastroenteritis.

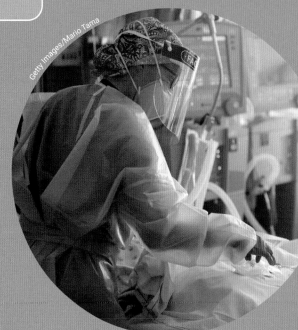

Getty Images/Mario Tama

▲ Medical staff in intensive care treating a patient

SA Health

Stop the spread of GASTRO

Stay home while unwell and do not return to work or school until no diarrhoea or vomiting for 48 hours.

Avoid preparing food for others for at least 24 hours if you have been sick.

Wash hands with soap and running water.

Wipe down surfaces.

Government of South Australia
SA Health

Reproduction

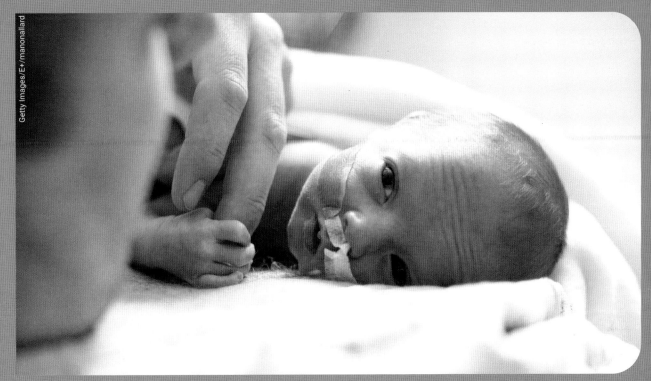

Getty Images/E+/manonallard

 FIGURE 3.0.1   Is vaping safe for developing babies? Studies show that e-cigarettes can affect baby brain development and are associated with low birth weight.

Have you seen people vape? Vaping involves heating a liquid with an electronic device and inhaling the vapour. Some people vape because they think it is less harmful and cheaper than smoking. Vapes contain several chemicals and toxins, including heavy metals. Even vape liquid that doesn't contain nicotine can still be very harmful. The vapours have been shown to slow brain development in teenagers and affect brain development of unborn babies. The World Health Organization (WHO) has issued warnings against their use.

▶ Why do you think some people choose to vape?

▶ Why do you think it is illegal to sell vape fluid containing nicotine in Australia?

#3 SCIENCE CHALLENGE ACCEPTED!

At the end of this chapter, you can complete the Big Science Challenge Project #3. You can use the information you learn in this chapter to complete the project.

Assessments
- Prior knowledge quiz
- Chapter review questions
- End-of-chapter test
- Portfolio assessment task: Science research and analysis project

Videos
- Science skills in a minute: Dissection **(3.10)**
- Video activities: Sexual reproduction in plants **(3.7)**; Asexual reproduction in plants **(3.7)**; Plants and pollinators **(3.8)**; AI in health **(3.9)**

Science skills resources
- Science skills in practice: Dissection skills **(3.10)**
- Extra science investigations: Growing plants from other plants **(3.7)**; Flower structure and function **(3.7)**; Looking at seeds **(3.8)**

Interactive resources
- Label: Binary fission **(3.1)**; Steps in meiosis **(3.2)**; Female reproductive system **(3.3)**; Male reproductive system **(3.4)**; Male and female flower parts **(3.7)**
- Crossword: Animal sexual reproduction **(3.5)**
- Drag and drop: IVF process **(3.6)**

To access these resources and many more, visit:
cengage.com.au/nelsonmindtap

Nelson MindTap

3.1 Asexual reproduction

BY THE END OF THIS MODULE, YOU WILL BE ABLE TO:

✓ define asexual reproduction, mitosis, binary fission, budding, fragmentation and parthenogenesis

✓ describe several different methods of asexual reproduction and give examples of each type

✓ explain why offspring produced by asexual reproduction are generally identical and why variation is only possible through mutation.

Interactive resource
Label: Binary fission

Quiz
Types of asexual reproduction

GET THINKING

Species need to reproduce to continue to produce new generations and survive. Can you name other methods of reproduction in addition to the method used by humans? Have you heard of binary fission, budding, fragmentation and parthenogenesis?

Reproduction

Are there any organisms that can live forever? Living things need a process that helps their species continue to exist. Reproduction is that process.

Methods of reproduction

When a living thing reproduces, it is known as the 'parent'. The new organism produced through reproduction of one or two parents is the **offspring**. When the offspring is only a cell, it may be called a 'daughter' cell. There are two main types of reproduction: sexual and asexual.

offspring
a new organism produced by asexual or sexual reproduction

Getty Images/Ian Waldie

▲ FIGURE 3.1.1 A koala parent with offspring joey, which was produced by sexual reproduction

For example, two koala parents produce offspring, a koala joey (Figure 3.1.1), by sexual reproduction, whereas bacteria, honey bees and some lizards produce offspring by asexual reproduction.

Asexual reproduction

What do bacteria and sea cucumbers have in common? They produce offspring by **asexual reproduction**. This is a method of reproduction that involves a single parent making a copy of itself to form offspring identical to the parent and each other.

asexual reproduction
a method of reproduction that involves one individual producing an identical copy of itself

mutation
a permanent change in the DNA in the cell of an organism

Asexual reproduction differs from sexual reproduction. Asexual reproduction is a relatively simple and fast method of producing offspring. There is no need to find a mate or use energy to transfer sperm to an egg. In asexual reproduction, there is only one parent, whose genetic material is passed to its offspring. As a result there is no genetic variation in asexual reproduction unless there is an error in the cloning process. This would result in the offspring having different genetic material (DNA). If it is a permanent change, this error is known as a **mutation**. Mutations can be passed on to offspring. Mutated DNA can cause a change in an individual's characteristics.

9780170472838

For example, mutations in some species of bacteria have given them the ability to resist some antibiotic medications.

In this module, we will look at four types of asexual reproduction: binary fission, budding, fragmentation and parthenogenesis. We will look at asexual reproduction in plants in Modules 3.7 and 3.8.

Binary fission

Simple organisms such as bacteria and paramecia (single-celled protists) use a simple form of asexual reproduction called **binary fission**. After a period of growth, the parent cell splits into two identical daughter cells – the offspring.

The main steps of binary fission are:

- replication – the cell duplicates everything, including DNA
- elongation – the cell grows longer
- fission – one parent cell splits into two identical daughter cells (Figure 3.1.2).

binary fission
a method of asexual reproduction in bacteria in which a parent cell splits into two identical daughter cells

mitosis
cell division and a form of asexual reproduction in some simple organisms

budding
a method of asexual reproduction in which a bud (a growth on the parent body) forms, grows and then separates or spreads

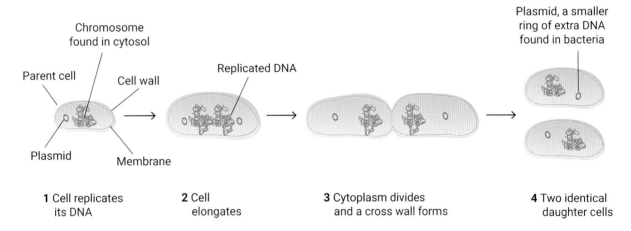

▲ **FIGURE 3.1.2** The main steps of binary fission in bacteria

Bacteria are prokaryotes – they do not have a nucleus. Simple eukaryotes – organisms with a nucleus – can also reproduce by binary fission. However, this can involve **mitosis**, in which a nucleus divides into two nuclei, then the cell divides.

Budding

Budding is a method of asexual reproduction in which a bud forms, grows and then separates (or spreads, in the case of coral) to form new offspring. When it is time to reproduce, a growth forms on the outer layer of the parent body. This growth is called a bud and grows by mitosis (cell division) until it develops into a tiny offspring. When it is mature, it breaks off to become an independent offspring. Small freshwater organisms called hydra, which grow to 1–2 cm long, use budding as their method of asexual reproduction (Figure 3.1.3).

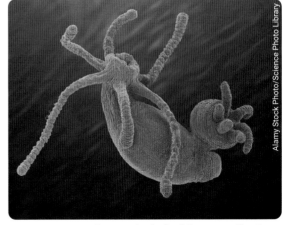

▲ **FIGURE 3.1.3** A parent hydra budding new offspring

Fragmentation

Fragmentation occurs when a body part or a fragment from a parent body breaks off to form a new independent offspring. For example, sea stars sometimes reproduce by fragmentation, especially when injured. When a fragment of a sea star breaks off, it develops into a new sea star that is genetically identical to the parent (Figure 3.1.4).

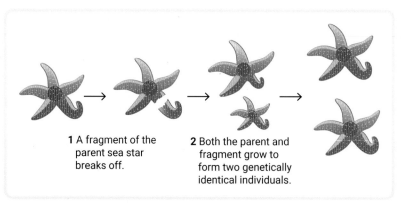

1 A fragment of the parent sea star breaks off.

2 Both the parent and fragment grow to form two genetically identical individuals.

▲ **FIGURE 3.1.4**   A fragment breaks off a sea star and both the parent and fragment grow to form two individuals.

Parthenogenesis

Parthenogenesis is the spontaneous development of an embryo from an unfertilised egg cell. A small but diverse range of animals reproduce by parthenogenesis. Some animals can even alternate between this form of asexual reproduction and sexual reproduction. Bees, ants, wasps and aphids and some lizards, birds and fish can all use parthenogenesis to produce identical young.

Did you know that queen honey bees can use parthenogenesis to produce multiple male offspring, called drones, for the sole purpose of mating? There is no need for sperm. The queen allows unfertilised eggs to mature, which develop into drones. Unlike worker bees, drones can have sex with the queen bee. As soon as the drones have finished fertilising eggs for the queen, they die. In comparison, female honey bee offspring are produced by an egg fertilised by sperm (sexual reproduction) (Figure 3.1.5).

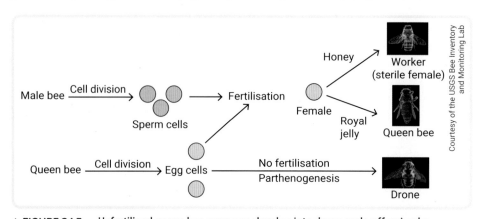

Courtesy of the USGS Bee Inventory and Monitoring Lab

▲ **FIGURE 3.1.5**   Unfertilised queen bee eggs can develop into drone male offspring by parthenogenesis.

Studies show that animals that reproduce by parthenogenesis can live in harsher environments than animals who only reproduce sexually. This is the case in regions with low rainfall where some animals, such as the Australian dwarf skink (Figure 3.1.6), reproduce through parthenogenesis. They do this to save energy and avoid dehydration. All-female **populations** of these lizards can reproduce without the need for a male.

population
a group of individuals of the same species that live in the same location

▲ **FIGURE 3.1.6**   Australian dwarf skink females can reproduce by parthenogenesis.

3.1 LEARNING CHECK

1 **Define** 'asexual reproduction'.

2 **Conduct** some research and then **state** the method of asexual reproduction for the following organisms.

   a Stick insect

   b Paramecium (Figure 3.1.7a)

   c Jellyfish

   d Planarian (Figure 3.1.7b)

   e Yeast

   f *Escherichia coli*

3 **Construct** a summary table for the four methods of asexual reproduction in this module: binary fission, budding, fragmentation and parthenogenesis. Use the column headings 'Method of asexual reproduction' and 'Description'.

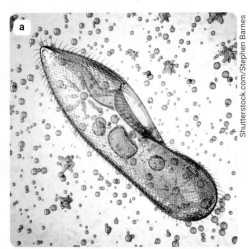

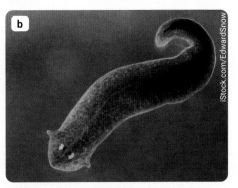

▲ **FIGURE 3.1.7**   (a) A paramecium and (b) a planarian

Chapter 3 | Reproduction **73**

3.2 Sexual reproduction

BY THE END OF THIS MODULE, YOU WILL BE ABLE TO:
✓ define sexual reproduction, gamete, meiosis, fertilisation, ova, gonad, sperm and zygote
✓ describe the method of sexual reproduction and explain how it results in genetic variations in offspring
✓ compare sexual and asexual reproduction.

Interactive resource
Label: Steps in meiosis

GET THINKING

Have you noticed that we all look very different from each other? Or did you know that dingos can have golden yellow, dark tan or even black fur? Diversity within a species is possible because of several processes associated with sexual reproduction. Can you think of what some of these processes might be?

Sexual reproduction versus asexual reproduction

sexual reproduction
a method of reproduction that involves two parents producing offspring that are not identical to the parent or each other (except twins)

Asexual reproduction requires one parent and results in identical offspring. In comparison, **sexual reproduction** requires two parents because two sets of genetic information are mixed, as shown in Figure 3.2.1. The offspring are not identical to the parents or to each other. Sexual reproduction requires more energy than asexual reproduction because parents need to travel to find partners and energy is required to mate.

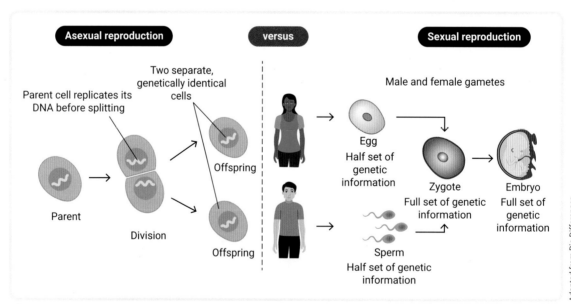

▲ **FIGURE 3.2.1** The difference between asexual and sexual reproduction

Sexual reproduction and variation

gamete
a sex cell of a sexually reproducing organism

meiosis
cell division that produces four non-identical gametes with half the genetic material of the parent cell

The most important benefit of sexual reproduction is genetic variation.

We look different from our parents and siblings because of genetic variation. Genetic variation is due to a mix of genetic material from each parent when the reproductive cells (**gametes**) join. Additionally, the process of making gametes, called **meiosis**, results in more genetic variation.

9780170472838

Meiosis

Meiosis is a process of cell division that produces gametes. At the end of the process, cells contain half the genetic material of the parent cell. The cell is then ready for sexual reproduction. Meiosis takes place in the **gonads** – male and female reproductive organs. Animals and plants both have gonads. Plants have their gonads on display, in flowers. Animals, such as mammals, have hidden gonads. Testicles, or testes, which produce sperm, are hidden inside the scrotum. The eggs or ova (singular: **ovum**) of mammals, are hidden inside ovaries (Figure 3.2.2).

gonad
a male or female reproductive organ such as the testicles and ovaries

ovum
a female gamete, or egg, produced in the ovary of a female

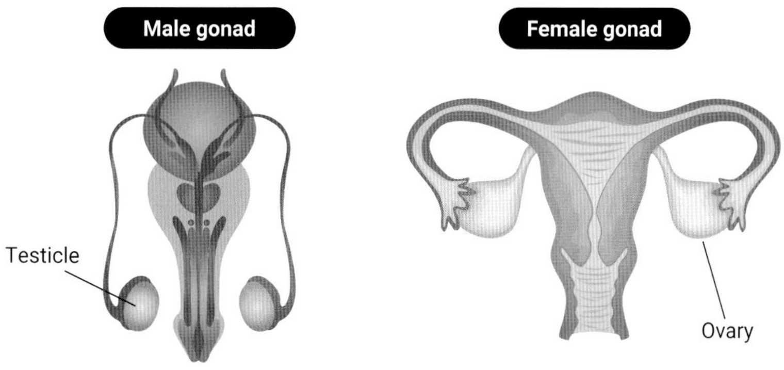

Male gonad | **Female gonad**

Testicle

Ovary

▲ **FIGURE 3.2.2**   The location of male and female gonads in humans

Prior to meiosis, DNA in the cells is duplicated in a process called DNA replication. During meiosis, the original cell divides twice to form two cells and then four cells (Figure 3.2.3). As part of this process, genetic material from both parent cells is:

- swapped and recombined in a process called **crossing over**
- further mixed up and shared in a process called **random assortment**.

The final four cells are non-identical gametes ready to be used in sexual reproduction. With only half the genetic material of the parent cell, they are ready to fuse with another gamete during fertilisation to form a cell with the full amount of genetic material.

crossing over
swapping of small amounts of maternal and paternal genetic material; happens before random assortment

random assortment
random recombining of large amounts of maternal and paternal genetic material; happens after crossing over

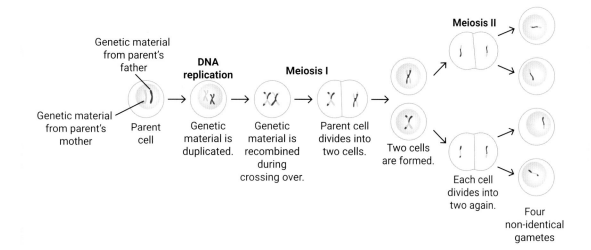

▲ **FIGURE 3.2.3**   The steps of DNA replication and meiosis

Fertilisation

zygote
the cell that results from the joining of a sperm and an egg cell at the completion of fertilisation

fertilisation
the process in which a male reproductive cell and a female reproductive cell (gametes) from each parent join to make the offspring's first cell, a zygote

sperm
a male gamete produced in the testes of male animals, or inside pollen grains of plants

When a male gamete and female gamete join to make the first cell of an offspring, the cell is known as a **zygote**. The fusion of the two gametes is called **fertilisation**. In mammals, such as humans, when **sperm** enters the vagina, it moves to the fallopian tube where it meets the egg, which has travelled the short distance from the ovary along the fallopian tube (Figure 3.2.4). The sperm's tail allows it to swim well in liquid. When the tail flicks, it propels the sperm forward. Sperm gain the energy to move long distances from their many mitochondria. Sperm use the energy to move along from the vagina through the cervix, uterus and fallopian tube, colliding with the egg midway along the fallopian tube. The egg's large structure stores nutrients, ready to sustain the growing embryo for a short time.

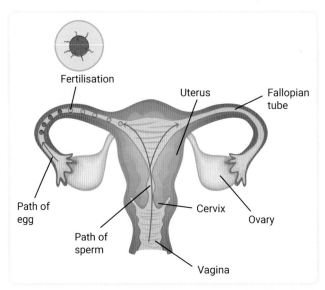

▲ **FIGURE 3.2.4**   The journey of sperm to the point of fertilisation

Have you seen the start of a marathon race? More than 5000 athletes start the Gold Coast and Margaret River marathons. Just like sperm in the fertilisation race, there can only be one winner. However, unlike the few thousand humans in a marathon race, hundreds of millions of sperm participate in the race to reach the egg (Figure 3.2.5).

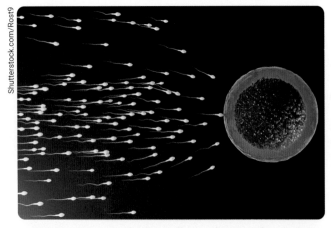

▲ **FIGURE 3.2.5**   Fertilisation will occur between the winning sperm and an egg.

Internal fertilisation

Organisms that live on land can lose water through evaporation and become dry and dehydrated. This is why fertilisation in most land animals occurs inside the animal – to protect the cells from the harmful dry environmental conditions. Animals that use internal fertilisation include most mammals and some reptiles.

9780170472838

External fertilisation

Water-based animals release their eggs directly into the environment. However, the eggs still need fertilising. In water, many animals release sperm to fertilise the water-suspended eggs. Once a sperm fuses with an egg, a zygote will form that develops into an embryo. Animals such as fish, frogs and mosquitoes fertilise eggs externally in water with little risk of the developing embryo drying out. Multiple eggs and sperm are released and fertilised. Frogs make sure that the sperm reach the eggs by the male climbing onto the female's back (Figure 3.2.6). This ensures the gametes are released at the same time and in a similar location, and therefore increases the chance of successful fertilisation. However, often there is a low survival rate of offspring fertilised externally because parents don't protect their offspring from predators. To counter this risk, a high number of gametes are released to ensure some offspring survive.

▲ **FIGURE 3.2.6** Frogs reproduce sexually by releasing eggs and sperm into a body of water for external fertilisation.

3.2 LEARNING CHECK

1 **Define** 'sexual reproduction'.

2 **Copy and complete** the following table comparing sexual and asexual reproduction in animals.

Factor	Asexual	Sexual
Number of parents		
Are offspring identical to parent?		
Type of cell divisions		
Are gametes needed?		
Three examples of organisms that use this method		
Relative speed of reproduction		

3 **State** the main advantage of sexual reproduction and **explain** why it is advantageous.

4 Choose two organisms from the following list.
 • Komodo dragon
 • Yeast
 • Coral
 • Squid
 • Aphid
 • Echidna

 a For each chosen organism, **predict** its method of reproduction as either sexual or asexual.

 b If the method of reproduction is asexual, **predict** the type of asexual reproduction.

 c **Research** facts about the two organisms and see if your predictions were correct.

BY THE END OF THIS MODULE, YOU WILL BE ABLE TO:

✓ define ovary, fallopian tube, uterus, cervix, vagina and ovulation

✓ label the key female reproductive structures on a diagram

✓ describe the function of the key female reproductive structures involved in ovulation, fertilisation and pregnancy.

Interactive resource
Label: Female reproductive system

GET THINKING

The structures of the female reproductive system are highly organised to perform special functions. They work in unison to ovulate, and when the opportunity arises, to facilitate fertilisation and pregnancy. You may be familiar with some structures already. As you read the module, take note of the less familiar structures and produce a table listing the terms and definitions.

Female reproductive structures

uterus
the organ where a fertilised egg implants and develops into an embryo and foetus

fallopian tube
a long narrow tube that connects one ovary to the uterus and is the site of fertilisation; sometimes called the oviduct

ovary (animal)
the female gonad where the ova (female gametes/eggs) are produced

ovulation
the release of an ova from an ovary

The main parts of the female reproductive system are the vagina, cervix, **uterus**, **fallopian tubes** and **ovaries** (Figure 3.3.1). Ova are produced in the ovaries. Did you know that the gametes of female babies develop in the uterus? Unlike sperm production in males, females produce all their ova prior to birth. That's a lot of meiosis in a short period of time!

On average, once a girl reaches puberty, an ovum is released from the ovary every 28 days, ready for fertilisation by sperm, or, if this does not occur, menstruation. Menstruation occurs when the uterus sheds its lining in the absence of fertilisation of the ova, commonly known as a 'period'. The two ovaries release ova alternately and the process is called **ovulation**. The ova travel through the fallopian tube towards the uterus. If sexual intercourse occurs within 5 days of ovulation, the ova may meet sperm, and fertilisation may occur.

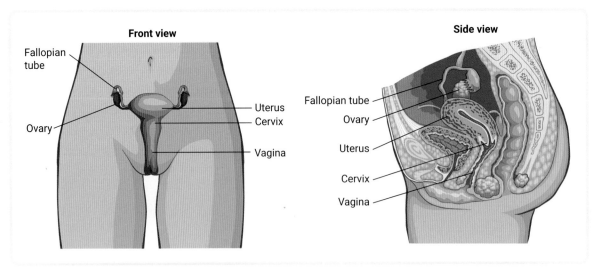

▲ **FIGURE 3.3.1** The human female reproductive system

Functions of the female reproductive system

Ovulation and fertilisation are two important functions of the female reproductive system. They are the preliminary processes prior to a pregnancy. When a sperm fertilises an ovum, the resulting cell has a full set of genetic material that is a combination of the sperm and ova's genes. The new cell is called a zygote. The zygote is transported towards the uterus. Along the way, it has already begun to divide into more cells by mitosis. The resulting ball of cells implants into the uterus and is now called an embryo.

In the second to eighth week of development, the unborn baby is called an embryo. After this time, it is called a foetus. In the fourth month, the foetus weighs about 110 g and its internal organs develop, including its reproductive organs (Figure 3.3.2). Once it is born, the foetus is known as a baby.

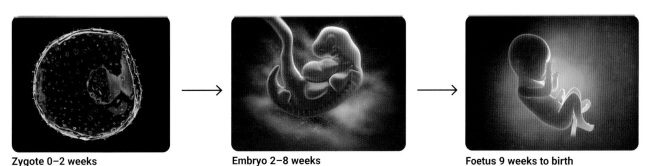

Zygote 0–2 weeks   **Embryo 2–8 weeks**   **Foetus 9 weeks to birth**

▲ FIGURE 3.3.2   The development of zygote to embryo to foetus

The **cervix** has important functions, including that it:

- allows sperm to pass from the vagina to the uterus
- allows menstrual flow to leave the uterus
- keeps the foetus in the uterus until it is time for birth, at which point the cervix opens to let the baby enter the vagina
- stops foreign objects entering the uterus, such as tampons and bath water.

The **vagina** receives the penis and the sperm that is ejaculated during sexual intercourse. The vagina also acts as the birth canal when a baby is born.

cervix
a muscular opening between the vagina and uterus

vagina
a muscular opening that leads from the external environment to the cervix

 **LEARNING CHECK**

1 **Describe** the process of ovulation.
2 **Copy and complete** the following table summarising the functions of the main parts of the female reproductive system.

Structure	Function
Ovary	
Fallopian tube	
Uterus	
Cervix	

3 Draw and **label** a diagram of the female human reproductive system (front view). Include the five main structures mentioned in this module. Indicate on the diagram where fertilisation takes place.
4 **Explain** why menstruation usually only occurs when an ovum isn't fertilised.

✓ define sperm, testes, scrotum, epididymis, vas deferens, urethra, penis, semen, prostate gland, seminal vesicles and bulbourethral gland

✓ label the key male reproductive structures on a diagram

✓ describe the function of the key human male reproductive structures.

Interactive resource
Label: Male
reproductive system

GET THINKING

The structures of the male reproductive system are highly organised to perform special functions. Each part has its own role to play within the whole system to enable reproduction. As you read this module, make a note of the differences in male parts compared with the female parts you learned about Module 3.3, and think about how the parts relate to their special functions.

testes
the male gonads that produce sperm (male gametes/sex cells)

scrotum
skin in the shape of a sac that wraps around the testes to keep sperm cooler than normal body temperature

Human male reproductive parts and their functions

The male reproductive system is responsible for sexual activity, the production of sperm and transfer of sperm to the female reproductive system, as well as urination.

Sperm (gametes) are produced in two testicles (**testes**), which are found inside a protective **scrotum**. Sperm need to be kept cooler than the internal body temperature, ideally at about 34°C (3°C below normal body temperature). This is why the scrotum is located outside of the body, as shown in Figure 3.4.1.

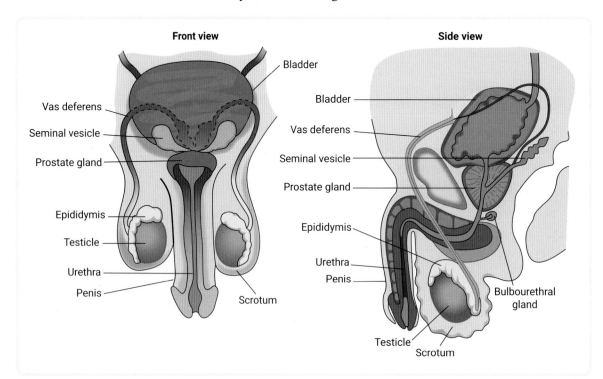

▲ **FIGURE 3.4.1** The human male reproductive system

When sperm is first produced, it is immature. The immature sperm are sent to the **epididymis**, a tube situated on the back of each testicle, to mature and be stored. During sexual arousal, muscle contractions force the sperm into the **vas deferens**, which are situated in the internal part of the body just behind the bladder. The vas deferens is a long muscular tube that transports the sperm to the **urethra**.

Sperm does not travel by itself. Along the way to the **penis**, fluid from the **seminal vesicle** mixes with the sperm and provides nutrients. Part of the urethra tube runs through the **prostate gland**. More nutritious fluid is added from the prostate gland, nourishing the sperm as it travels. Just below the prostate gland and located on both sides of the urethra are the **bulbourethral glands**. These glands secrete a lubricating fluid for the urethra that makes the urethra slippery for the smooth flow of sperm. It also neutralises any acidic leftover urine.

Sexual intercourse for fertilisation

The urethra has the dual purpose of transporting urine from the bladder outside of the body through the penis and enabling the ejaculation of sperm. The process of ejaculation occurs when the male is sexually aroused. For the purpose of reproduction, the sperm is ejaculated into the female's vagina. Before ejaculation, when the penis is erect and enlarged due to a surge of extra blood, the flow of urine is blocked. Only **semen** (the mix of sperm and fluids) can be ejaculated at this time.

3.4 LEARNING CHECK

1 **Describe** the process of ejaculation for the purpose of sexual reproduction.
2 **Copy and complete** the following table summarising the functions of the main parts of the male human reproductive system.

Structure	Function
Testes	
Scrotum	
Epididymis	
Vas deferens	
Urethra	

3 **Describe** what semen is and how it is produced. Include the terms 'prostate gland', 'seminal vesicles' and 'bulbourethral gland' in your answer.
4 Draw and **label** a diagram of the male human reproductive system (side view).

epididymis
a tube that sits on the back of each testicle; stores sperm

vas deferens
a long, muscular tube that transports sperm from the epididymis to the urethra

urethra
a tube that transports urine and semen, extending from the bladder to the tip of the penis

penis
a sex organ that inserts into the vagina to transfer sperm into the female reproductive system

seminal vesicle
one of two sac-like glands that secrete fluids that are the main component of semen

prostate gland
a gland that produces fluids found in semen, located between the bladder and the penis

bulbourethral gland
one of two glands located on both sides of the urethra, below the prostate gland, that secrete a lubricating fluid

semen
a mix of sperm and fluids that provide nutrition and lubrication for sperm

3.4

✓ define gestation

✓ relate offspring survival rate to the quality of parental care

✓ relate an animal's type of reproductive strategy to its habitat.

Interactive resource
Crossword: Animal sexual reproduction

GET THINKING

After birth, offspring of different species receive different amounts of parental care. As a human baby, you were cared for by your parents or carers, but some species are left to fend for themselves immediately after the sperm and eggs are released. Get ready to find out about a small sample of diverse reproductive methods used by some animals.

Gestation in humans and other animals

Can you think of your 'happy place' where it is warm, snug and soothing? The uterus is that kind of place for an embryo and foetus. As soon as fertilisation occurs, the zygote moves through the fallopian tube and implants into the lining of the uterus.

This implantation of is crucial for the development of the embryo.

A developing human baby remains in the uterus for about 40 weeks. This is called the **gestation** period. The duration of gestation varies between different species. Figure 3.5.1 shows the gestation period for some animals. Can you guess the gestation periods for any other animals?

gestation
the period between fertilisation and birth, during which a foetus (unborn baby) develops

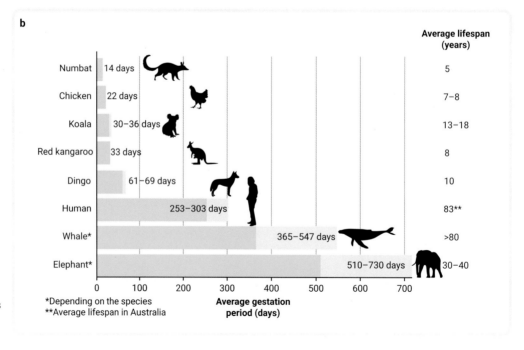

Animal	Average gestation period (days)	Average lifespan (years)
Numbat	14 days	5
Chicken	22 days	7–8
Koala	30–36 days	13–18
Red kangaroo	33 days	8
Dingo	61–69 days	10
Human	253–303 days	83**
Whale*	365–547 days	>80
Elephant*	510–730 days	30–40

*Depending on the species
**Average lifespan in Australia

▶ **FIGURE 3.5.1**
(a) The gestation time for humans is 40 weeks.
(b) Different animals have different gestation periods.

Other reproductive methods related to habitat

Animals that inhabit dry land perform internal fertilisation as their reproductive method. The fusion of gametes takes place inside the female. This method protects gametes from the effects of the environment such as extreme temperatures, and predators, and increases the chance of survival until **birth**. Although internal fertilisation is a great survival mechanism for all land-dwelling animals, the way they produce offspring can differ.

birth
the process in which the foetus is pushed out of the uterus and leaves the mother's body

Birds, such as chickens, lay fertilised eggs. Marsupials, such as kangaroos, give birth to under-developed joeys who finish developing in the pouch where they can attach to a nipple. Other mammals, such as dogs, give birth to well-developed pups (Figure 3.5.2). Reptiles lay eggs that receive nourishment from the yolk.

Some animals that live in water also use internal fertilisation, such as dolphins, who are closely related to all other mammals. Other aquatic animals use external fertilisation. Eggs are released into the water so they can be fertilised by a male's swimming sperm. Animals that use external fertilisation need water for the sperm to swim towards the egg. These animals also need water so that the gametes don't dry out.

▲ **FIGURE 3.5.2**　French bulldogs give birth to well-developed offspring.

The reproductive method of seahorses is particularly interesting. Female seahorses deposit up to 1500 eggs into the male's pouch. The male's sperm is released into the pouch to fertilise the eggs. The pregnancy continues for up to 4 weeks inside the pouch of the male where oxygen and nutrients are supplied to the embryos (Figure 3.5.3).

▲ **FIGURE 3.5.3**　**(a)** Seahorses dance before the female deposits her eggs into the male pouch for fertilisation. **(b)** The male seahorse carries the embryos.

▲ **FIGURE 3.5.4**   Coral reproduce by spawning.

spawning
the release of male and female gametes into the environment (usually aquatic), during which sperm randomly fertilise eggs

Spawning is another method of external reproduction. Spawning occurs when both eggs and sperm are released into the environment. Coral is an animal that reproduces through spawning. When coral reproduce sexually, male and female gametes are released into the water at the same time (Figure 3.5.4). Along the Queensland coast, sperm and eggs are released annually by coral in the Great Barrier Reef. Around the full moon in November, scuba divers film this amazing phenomenon. When a sperm randomly fertilises an egg, the fertilised egg sinks to the bottom of the ocean floor, ready to grow into a new coral. Nearly 4000 km away, at Australia's other famous reef, Ningaloo Reef on the coast of Western Australia, the waters become cloudy with the release of coral sperm and eggs in March–April each year. Many ocean species, such as fish, use spawning as their method of reproduction. The aquatic habitat allows for spawn to be released and travel for a period of time without the drying effects of a land-based habitat.

Parental care and survival rate

Once offspring are born or spawn are released, they receive different levels of parental care, depending on the species and habitat. Offspring that receive more parental care or input have better chances of survival. When fish release their eggs and sperm, that marks the end of their parental care. When horses are born, they are walking within an hour, unlike human babies who normally take around a year to do so. Foals tend to become independent of parental care by the fourth month. Some human babies do not leave home until well after 18 years!

Feeding and protecting offspring can be hard work and requires an input of energy from the parents. You may have observed birds soaring towards a nest and dropping off prey. You may have been a victim of a swooping bird as it protects its young. These are both examples of parental care.

Parental care can take different forms. Birds build nests and deliver food until the chicks leave the nest after a few weeks (Figure 3.5.5). Mammals feed their offspring milk. Numbats teach their young how to hide from predator birds, and many dingoes are taught how to hunt for prey by their parents. There is a higher chance of survival if offspring are fed, protected and taught how to be independent.

Some organisms have fewer offspring and invest much energy into their care, whereas other organisms use a lot of energy to produce many gametes, and potentially many offspring, but invest little energy into their care. These are both survival mechanisms that ensure offspring survive to adulthood and can themselves then reproduce.

▲ FIGURE 3.5.5 An Australian magpie-lark parent feeds its chicks until the chicks are ready to leave the nest after a few weeks

3.5 LEARNING CHECK

1 **Define** 'gestation'.

2 Find out what your gestation period was. Were you born premature (before 37 weeks) or later than 40 weeks? Survey 3–5 classmates to find out their gestation periods, and then **calculate** an average.

3 **Research** the gestation period of three other animals not mentioned in this module.

4 Emperor penguin chicks have a high mortality rate because of predation and starvation. Males and females share the parental care. **Analyse** the infographic in Figure 3.5.6 to describe the steps taken by both parents to care for their young.

▲ FIGURE 3.5.6 Male and female Antarctic emperor penguins share the parental care.

5 **Describe** the parental care of one species of animal that reproduces sexually, that was not mentioned in the module.

BY THE END OF THIS MODULE, YOU WILL BE ABLE TO:

✓ define infertility and in vitro fertilisation, and explore related ethical considerations

✓ describe the different reasons for infertility

✓ describe some of the medical and scientific methods that can be used to help humans reproduce.

Interactive resource
Drag and drop: IVF process

GET THINKING

Some people find reproducing easy, whereas others struggle to conceive (start a pregnancy). For these people, infertility can be a cause of great distress and sadness. In vitro fertilisation (IVF) and related treatments have assisted many people in their quest to have children. You will be asked to think about the scientific facts and the ethics to help you consider what is right and wrong for all parties involved in this treatment.

Infertility

If a person or a couple have been trying to get pregnant for one or more years without success, doctors may classify them as **infertile**. If you recall the steps to pregnancy, you may already be able to predict some of the causes of infertility.

To diagnose and then treat the specific cause of infertility, doctors will order a series of tests to analyse possible causes in both the male and female.

infertile
a term used when a female has been unable to get pregnant after one year of trying

endometriosis
a painful disorder experienced by some women when tissue similar to uterus lining grows on the outside of the uterus

▼ **TABLE 3.6.1** Different causes of infertility

	Some causes of infertility
Men	• No sperm • Low sperm count • Poor sperm motility (movement) • Low hormone levels
Women	• Blocked fallopian tubes • Low or high hormone levels • Endometriosis • Cysts on ovaries • Conditions that affect ovulation, such as polycystic ovary syndrome

In males, the number and quality of sperm can be affected by many things, including overheated testicles, testicle trauma, heavy alcohol use, drug use or smoking, or a genetic disorder. Causes of female infertility can include a reduced number of eggs, cysts on the ovaries, hormonal imbalance and menopause (the stage of a woman's life when she has finished ovulating). Smoking and alcohol consumption are also contributors, and some diseases, such as chlamydia, can result in blocked fallopian tubes.

Depending on the test results, some infertile people will use reproductive technologies known as **in vitro fertilisation (IVF)**. Some people without fertility issues also use IVF to become pregnant, such as same-sex couples, people without partners, or people who want to avoid passing on a genetic disorder.

in vitro fertilisation (IVF)
a complex series of procedures that involve the fertilisation of eggs in a laboratory and implanting the embryo (fertilised egg) in the uterus

IVF and other treatments

IVF treatment involves procedures in which eggs are extracted from a female and fertilised in a laboratory, and then the embryo is placed in the female's uterus (Figure 3.6.1). New rapid-freezing technology allows doctors to pause the transfer of an embryo into the uterus until the optimum time in a female's menstruation cycle for implantation.

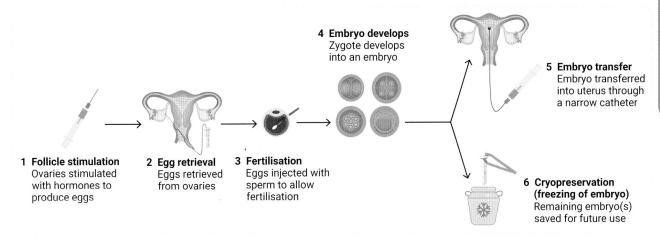

4 Embryo develops
Zygote develops into an embryo

5 Embryo transfer
Embryo transferred into uterus through a narrow catheter

1 Follicle stimulation
Ovaries stimulated with hormones to produce eggs

2 Egg retrieval
Eggs retrieved from ovaries

3 Fertilisation
Eggs injected with sperm to allow fertilisation

6 Cryopreservation (freezing of embryo)
Remaining embryo(s) saved for future use

▲ **FIGURE 3.6.1** The IVF process

The first IVF baby was born in 1978. According to the University of New South Wales, nearly one in 20 children was born through IVF treatment in 2018. In 2021, a total of 6071 people underwent assisted reproduction treatment in Western Australia.

Unfortunately, IVF treatment does not guarantee a baby. One of the main factors affecting the success rate of IVF treatments is the age of the parents.

Genetic testing and IVF

There are strict rules about the storage of embryos, testing of embryos, surgical procedures, and research involved in IVF and other assisted reproductive technologies. Some of these rules relate to the complicated **ethics** about using embryos. For example, in Australia the genetic testing of embryos produced through IVF is only allowed when there is a known risk of a serious genetic condition, such as haemophilia, Huntington's disease, sickle cell anaemia, cystic fibrosis and cancer resulting from the *BRCA1* gene mutation.

ethics
moral considerations that govern decisions or actions that can affect people or other organisms

Other innovative procedures are being used in conjunction with IVF treatment to prevent disease in offspring. For example, in Australia, a law was passed in 2021 that allows the donation of mitochondria. Mitochondrial donation is a procedure performed through an IVF treatment that replaces abnormal mitochondria in an ovum (egg) with healthy mitochondria from a donated egg. The technique is used to prevent serious mitochondrial disease, such as muscular dystrophy, being passed from mother to child.

Sometimes people use donor sperm or eggs to avoid passing on a serious genetic condition (Figure 3.6.2).

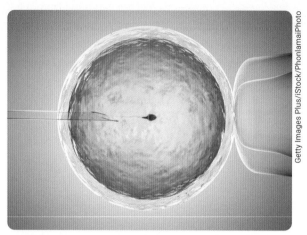

▲ FIGURE 3.6.2 Injecting donor sperm into an egg is one type of IVF procedure.

Access to donor records

Since 2005, the practice of anonymous egg and sperm donation in Australia no longer occurs. This practice was changed to give children conceived by egg or sperm donation greater transparency and information about their biological parents. This is important for many reasons, including:

- the option of knowing about or potentially meeting their donor and any genetic siblings
- knowing about any genetic diseases or conditions that they may have inherited from their donor.

The rules about how and when a child can access information about their donor can be complicated and vary between states and territories. This is because the ethics around access to donor records is complex, with many points of view and arguments that need to be considered. For example, do you think a child has a right to know who their biological parent is? What about if the donor does not want to be identified or involved in a child's life? And what about the rights of the people who are using donor eggs or sperm to conceive a child?

3.6 LEARNING CHECK

1 **Define** 'infertile'.
2 Conduct further research into mitochondrial disease. **Explain** why mitochondrial disease can cause severe symptoms.
3 **Construct** a Venn diagram for the causes of infertility in males and females.
4 **Define** 'ethics' and explain how it relates to IVF.
5 **Describe** the process of IVF.
6 **List** three ethical considerations that parents undergoing IVF treatment could make before selecting embryos.

9780170472838

GET THINKING

Have you kept potatoes for too long in your pantry? Have you seen how they started propagating with new green shoots (axillary buds)? How do you think this could relate to reproduction?

Sex and flowers

Flowering plants all use the same method of sexual reproduction to produce offspring. Flower structures have the sexual forms and functions that enable **pollination** and fertilisation and the production of non-identical offspring with genetic variation.

Male components of flowers

The male component of the flower is called the **stamen**. It is made up of an **anther** at the end of a **filament**. The filament's purpose is to elevate the anther above the flower to make it accessible to **pollinators**, such as bees, birds and bats. Male gametes (sperm) are produced and found inside **pollen** in the anther of a flower.

pollination
the transfer of pollen from a male anther to a female stigma

stamen
the male reproductive part of a flower, consisting of a filament and anther

anther
the section of the male part of a flower (stamen) that produces pollen grains

filament
a tall structure that elevates the anther so that it sticks out of the flower, making it visible to pollinators

Female components of flowers

Starting from the base of the flower, an **ovary** stores female gametes inside the **ovules**. You can follow this structure from the base, up the middle tube-like **style**, to the sticky **stigma** at the top. Its stickiness serves to catch swirling pollen in the air when they blow off the male anther or are carried to the flower by a pollinator such as a bird or a bee. The

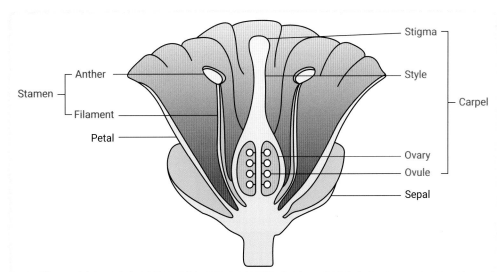

▲ **FIGURE 3.7.1** The structures of the reproductive system of a flower. The male parts are labelled in green; the female parts are labelled in orange.

entire set of female components, including stigma, style, ovary and ovules, make up the structure of the **carpel** (Figure 3.7.1). Fertilisation of a male and a female gamete occurs inside the ovule.

pollinator
an animal that transports pollen from the male anther to the female stigma for pollination

pollen
grains that produce and hold male gametes (sperm) on the anther of a flower

ovary (plant)
the structure at the base of a flower that encloses the ovules

ovules
structure inside the ovary that makes and contains female gametes

style
a long stalk that connects the ovary to the stigma

stigma
the sticky tip of the carpel where pollination occurs

carpel
the female reproductive part of a flower consisting of the stigma, style, ovary and ovules

petal
the colourful part of a flower that attracts pollinators

sepal
one of many structures that cover the flower (bud) as it forms

nectar
a sugary liquid in the middle of a flower, produced by the flower, which attracts pollinators

pollen tube
a tube that develops when a male gamete (inside pollen) lands on a stigma

seed
the structure that forms from the ovule after it is fertilised, usually found in fruit

Petals and sepals

The colourful parts of the flower are the **petals**. The role of petals in reproduction is to attract as many pollinators as possible to the eat from the flower's nectar and therefore pollinate the plant. **Sepals** look like leaves, but they are the special structures that cover the flower as it forms. When the flower is ready to bloom, the sepals open to reveal the flower.

Pollination and fertilisation

Pollination is the transfer of pollen from a male anther to a female stigma. Although the anther is generally very close to the female stigma, flowers usually can't pollinate themselves without some help. Flowers rely on wind or animal pollinators to transfer the pollen. For example, bees rub against an anther when feeding on **nectar**, picking up pollen. When a bee flies to another flower, it may rub this pollen onto the stigma (Figure 3.7.2). Once pollination has occurred, a **pollen tube** grows down through the style to the ovary to reach the ovule. Sperm from the pollen fertilises the gametes (ova) inside the ovule.

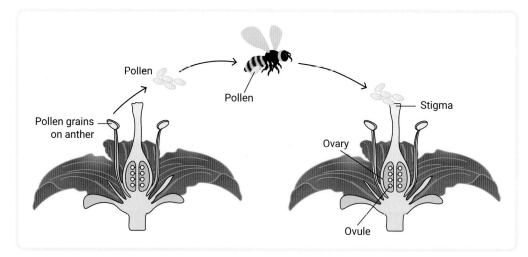

▲ **FIGURE 3.7.2**   In pollination, a bee transfers pollen from one flower's anther to another flower's stigma.

The fertilised ovule forms the **seed**, and the ovary becomes the fruit that usually envelopes the seed (Figure 3.7.3).

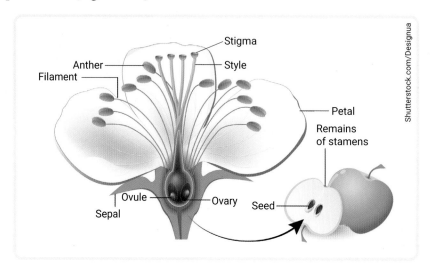

▶ **FIGURE 3.7.3**   After fertilisation, a fertilised ovule becomes the seed, and the ovary becomes the fruit.

9780170472838

Sexual versus asexual reproduction in plants

Plant reproduction that involves the fusion of male and female gametes results in offspring that are genetically different. This form of plant reproduction is sexual reproduction. Asexual reproduction in plants can occur too and is often called **vegetative reproduction** or propagation. This process involves new plants forming from vegetative parts of the original plant, such as the leaves, stems and roots (Figure 3.7.4).

vegetative reproduction
a method of asexual reproduction in which new plants form vegetative parts of the original plant, such as the leaves, stems and roots

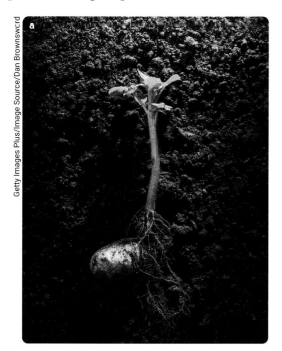

▲ **FIGURE 3.7.4** **(a)** Vegetables such as potatoes propagate from tubers, which are thick swollen parts of underground tissue. **(b)** Tulips and lilies propagate from bulbs, which are underground buds that store food for the plant.

Vegetative reproduction produces offspring that are identical to the parent plant. As a result, there is not much genetic variation from one generation to the next. This very low amount of variation can survive well if the environment is stable. Farmers take advantage of this when they grow crops with a desired trait, such as frost tolerance in strawberries or tomatoes. Reproducing crops by vegetative propagation (asexual reproduction) guarantees the desired trait will be passed to offspring.

Video activities
Sexual reproduction in plants
Asexual reproduction in plants

Interactive resource
Label: Male and female flower parts

Extra science investigation
Flower structure and function
Growing plants from other plants

3.7 LEARNING CHECK

1 **State** the names of the male and female organs of the flower.
2 Draw and **label** a diagram of the sexual parts of a flower.
3 **Compare and contrast** asexual and sexual reproduction in plants.
4 **Create** a crossword using the names of the sexual parts and functions of the flower.

Include clues for at least four words across and four words down. Swap with another student to find out if the crosswords work and simultaneously review what you learned.

3.8 Pollination and seed dispersal

BY THE END OF THIS MODULE, YOU WILL BE ABLE TO:

✓ state the types of pollination, recall the processes and list different pollinators

✓ describe different methods that increase pollination

✓ define seed dispersal and describe different methods of seed dispersal.

Video activity
Plants and pollinators

Extra science investigation
Looking at seeds

cross-pollination
the transfer of pollen grains from one flower to a flower on a different plant

self-pollination
the transfer of pollen within the same flower or plant

GET THINKING

Animals can transport and deliver seeds in their faeces to a new habitat after eating and digesting the fruit or seeds of a plant. In this module, you will find out other ways seeds are dispersed.

Pollination and pollinators

The transfer of pollen from an anther to a stigma is called pollination. Two types of pollination are:

- **cross-pollination** – the transfer of pollen grains from the anther of one flower to the stigma of another flower
- **self-pollination** – the transfer of pollen within the same flower (Figure 3.8.1).

Pollen can be transported to different plants but can only pollinate a plant of the same species. Who does the pollinating and how? In Australia, common pollinators are bees, flies, wasps, beetles, butterflies, moths, bats and birds. Wind can blow pollen from the anther to the stigma, so this is also regarded as a pollination mechanism.

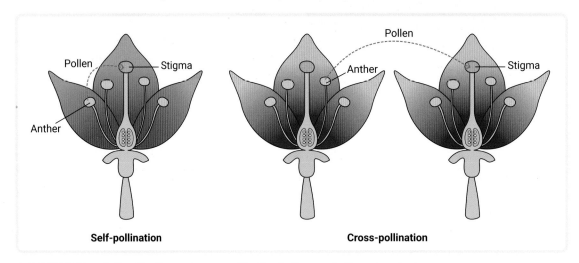

▲ **FIGURE 3.8.1** Self-pollination and cross-pollination both occur in flowering plants.

Methods that increase pollination

We depend on pollinators to pollinate our native plants and agricultural crops. In Australia, the rich diversity of insect and bird species consume the delicious nectar contained in flowers and transfer pollen from flower to flower (Figure 3.8.2).

Flower shape, scent and colour can assist in pollination. As pollinators are enticed into the flower, there is more chance of pollen being rubbed onto the surface of the pollinator and carried away to the next flower's stigma. Cycads from inland Australia use scent to attract thrips for pollination. Thrips are small insects that cannot carry many pollen grains, so a plant needs to attract large numbers of them.

However, some practices in society decrease the rate of pollination. For example, the use of agricultural chemicals, such as pesticides, can harm important pollinators such as insects, bees and beetles.

▲ **FIGURE 3.8.2** **(a)** Birds, insects and other animals are pollinators. **(b)** A bee brushes yellow pollen grains from one flower onto another flower as it feeds on nectar.

Seed dispersal

A plant seed contains an embryo and an energy store and is surrounded by a hard protective coating. Seeds can be different shapes and sizes. If you pick a dandelion and blow the white feather-like structures into the air to watch them spin around like parachutes, you are actually dispersing seeds, as shown in Figure 3.8.3. Seed dispersal is the transportation of seeds from the parent plant to a new place ready to germinate a new plant to grow and live.

▲ **FIGURE 3.8.3** Blowing dandelion seeds into the air disperses them.

Methods of seed dispersal include:

- wind, such as for dandelions
- animals consuming and digesting fruit, and expelling seeds in faeces, such as elephants eating acacia seeds
- animals carrying seeds within burrs in their fur or on their skin, such as bindies caught in dog fur or human clothing (Figure 3.8.4).

Seed dispersal can be beneficial for the plants as it prevents competition for food, sunlight and water between the parent and offspring.

▲ **FIGURE 3.8.4** This dog's fur has seeds stuck in it; this helps the seeds disperse.

3.8 LEARNING CHECK

1 **Describe** seed dispersal.
2 Emus are famous Australian birds that disperse seeds. **Research** two or three species of plant that emus disperse and **describe** the method of dispersal.
3 In terms of survival, **state** which type of pollination (self-pollination or cross-pollination) would be most advantageous and **explain** why.
4 **List** four different pollinators.

3.9 Assistive reproductive technologies

BY THE END OF THIS MODULE, YOU WILL BE ABLE TO:

✓ discuss how Australia has developed an artificial intelligence (AI) system that is used to predict the likelihood of a viable pregnancy from transfer of a single embryo to a woman undergoing IVF

✓ relate the needs of society to the focus of AI scientific research on embryo selection.

Video activity
AI in health

What is artificial intelligence?

Artificial intelligence (AI) can be described as the development of computer systems to perform tasks that would normally require human intelligence. AI is useful for efficiently solving complex problems by mimicking human intelligence.

Unlike coding, which is designed to perform a specific task, AI machines can perceive and predict, answer generic questions, learn, solve problems and design tasks (Figure 3.9.1).

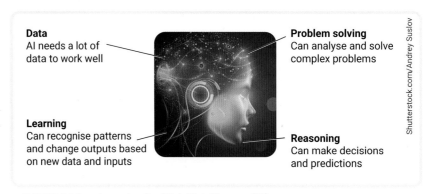

Data
AI needs a lot of data to work well

Problem solving
Can analyse and solve complex problems

Learning
Can recognise patterns and change outputs based on new data and inputs

Reasoning
Can make decisions and predictions

Shutterstock.com/Andrey Suslov

▲ **FIGURE 3.9.1** Aspects of artificial intelligence (AI)

How can AI improve IVF?

One of the first babies in Australia to be born with the help of artificial intelligence was Charlotte from the Sunshine Coast in 2020. When Charlotte was an embryo, AI was used to identify her as the embryo most likely to result in pregnancy and a healthy foetus out of a group of potential embryos. Charlotte's embryo was selected to be transferred to her mother's uterus.

Shutterstock.com/Krakenimages.com

▲ **FIGURE 3.9.2** AI can improve the accuracy of embryo selection, rather than just relying on the judgement of an embryologist.

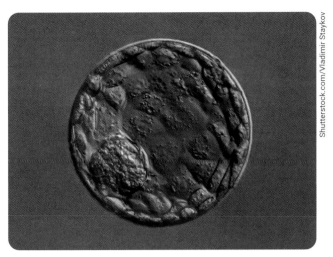

▲ **FIGURE 3.9.3** A 5-day old embryo – classified as a blastocyst, ready to be transferred to the uterus.

Without AI, embryologists select the embryo they think has the most potential. They do this by looking at it under the microscope and using a grading system to select the embryo with the highest potential for a successful pregnancy and birth.

In contrast, AI techniques for embryo selection rely on huge amounts of data in the form of hundreds of images of embryos. For example, AI uses embryo images showing the growth patterns that are related to whether each embryo developed into a successful pregnancy.

AI can assess the growth of the embryos over 5 days and then relate this data to whether a foetal heart has developed or not. In this way, AI helps in the decision-making around selecting an embryo with the highest potential for a successful pregnancy. A higher chance of survival can save patients time and money and potentially reduce the emotional stress involved in IVF.

3.9 LEARNING CHECK

1 **Describe** artificial intelligence (AI).
2 **Discuss** how AI is used in IVF.
3 Would you allow AI to choose your embryo? Or would you prefer an embryologist? **State** which you prefer and **explain** your reasoning.
4 **State** why AI is becoming the preferred method of embryo selection.

SCIENCE SKILLS IN FOCUS

IN THIS MODULE, YOU WILL FOCUS ON LEARNING AND IMPROVING THESE SKILLS:

▶ conducting a safe dissection of a flower

▶ drawing a labelled diagram of a dissected flower

▶ developing an explanatory model for vegetative reproduction.

Dissections involve carefully cutting open and exposing the interior part of a plant or an animal. They help you understand the relationship between internal structures and their functions.

Remember these rules when doing a dissection.

1 Wear safety glasses.

2 Use a dissecting board or tile underneath the newspaper.

3 Keep sharp tools on the table when not in use and do not walk around with them.

4 Return used tools upside down in a container of disinfectant.

Typical dissection tools and their functions are listed in Table 3.10.1.

▼ TABLE 3.10.1 Dissection tools

Tool	Function
Scissors	Cutting apart tissues (such as petals, stems, roots)
Forceps (large tweezers)	Holding a sample while cutting tissue
Mounted needle	Holding a sample in place
Scalpel	Making incisions
Dissection pins	Securing parts of a sample to the board

When you draw a dissected structure:

• use an informative title

• make your drawing large enough for your teacher to see all parts (half a page is usually big enough)

• draw with a pencil

• label structures.

INVESTIGATION 1: DISSECTION OF A FLOWER

AIM

To locate, draw and study the reproductive structures of a flower and relate them to their functions

MATERIALS

☑ assorted flowers (tulips, daffodils and gladiolus are ideal)

☑ scalpel

☑ scissors

☑ hand lens

☑ dissecting board and newspaper

☑ piece of A3 poster paper

☑ sticky tape

☑ stereomicroscope (optional)

METHOD

1 Hold one of the flowers and observe the external parts (refer to Figure 3.10.1). Name them as you point at them. Try to rub some pollen off the anther. Note the colour. If you have access to a stereomicroscope, view the anther under the microscope.

2 Carefully cut the flower in half lengthwise. Observe the internal organs.

3 Use a hand lens to magnify each structure. Study the structures carefully.

4 Draw a labelled diagram of the flower half.

5 Count the number of petals if there are not too many.

6 Use the scalpel to carefully remove the male and female parts. Arrange them on the poster paper and label each part. Refer to Figure 3.10.1 for guidance.

Video
Science skills in a minute: Dissection

Science skills resource
Science skills in practice: Dissection skills

7 If your class was given a variety of flowers, look at other students' dissections and note the differences in colour, length of stigma and any other variables you see.

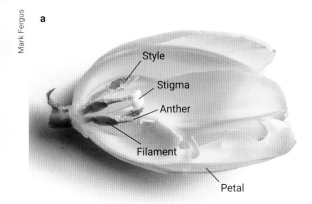

a

Style
Stigma
Anther
Filament
Petal

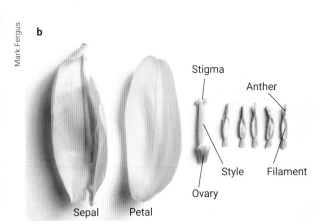

b

Stigma
Anther
Style   Filament
Ovary
Sepal   Petal

▲ FIGURE 3.10.1   (a) Flower dissection and (b) reproductive parts of the flower

EVALUATION

1 Explain why students and researchers need to do dissections.

2 List three observations that were different in the flowers of different species of plants. Use your senses of sight, smell and touch.

3 Name and describe three structures whose shape or features helped with their function.

4 State one feature of the reproductive organs you thought was most interesting and explain why.

5 Count the number of ovules – are there eight?

6 Recall where the female gametes (ova) are found. Describe the benefit of this location.

INVESTIGATION 2: INVESTIGATING VEGETATIVE REPRODUCTION IN SWEET POTATOES

AIM

To grow sweet potato offspring from one parent sweet potato tuber, and to investigate factors that promote or reduce growth by choosing one independent variable to test

HYPOTHESIS

Record your chosen independent variable and write a hypothesis.

Write a prediction based on your hypothesis

Note: A whole sweet potato takes 10–14 weeks to mature. The 'vegetable' part of the sweet potato is called a tuber, which is where roots and shoots grow from.

MATERIALS

☑ 2 sweet potatoes per group
☑ 2 beakers, one for each tuber
☑ 8 toothpicks
☑ water
☑ soil
☑ 2 plant pots
☑ slow-release fertiliser (optional)

METHOD

1 Locate the middle of each sweet potato and insert four toothpicks around the centre. This will help the sweet potato stand half in and half out of the beaker.

▲ FIGURE 3.10.2 Insert toothpicks to support the tuber in the beaker of water similar to this example shown.

2 Place the sweet potatoes in the beakers with the root end, which usually has tiny roots, at the bottom. Fill the beakers with enough water to cover the bottom half of the tuber.

3 Decide which variable you will change for your test; for example, sunlight, salinity or fertiliser. Write a hypothesis for your test sweet potato.

4 Place your control and test sweet potatoes in a sunny spot. The exception to this is if your independent variable is sunlight. In this case, place the test sweet potato inside a dark cardboard.

5 Change the water of both beakers every couple of days. Make sure the bottom of the tuber remains submerged in water.

6 Observe the sweet potato for a few weeks until green sprouts form. These are known as 'slips'.

7 Separate two or three slips from the sweet potato by gently twisting them.

8 Lay the slips in a shallow dish. Pour in enough water to submerge the stem of the slips, but let the leaves hang over the edge of the dish. New roots should begin to grow within a few days.

9 When the roots are 2–3 cm long, fill a plant pot with soil and place the slips halfway down, roots facing into the soil.

◀ FIGURE 3.10.3 Sweet potato slips

10 Care for your plants by giving them regular light, water and sunlight, continuing to maintain the same independent variable. If water is your independent variable, water your control plant but not your test plant. If sunlight is your independent variable, keep the test plant in a dark spot.

11 After a few weeks, transfer the pot outside and continue to care for the plants for 10–14 weeks.

RESULTS

1 After 10–14 weeks, check on your tuber offspring.

2 Measure your dependent variable, such as growth, and record your results.

EVALUATION

1 Record your results in a table.

2 Do your results support your hypothesis?

3 List two variables you controlled, in the control and test pots, so that you could produce valid results.

4 Name two other variables that you could have chosen as an independent variable and explain how they would have affected the growth of the tuber.

5 Describe a variable that was difficult to control and explain how it may have affected the validity of the results.

CONCLUSION

Look at the experiment aims, your hypothesis and your results to write a conclusion based on your hypothesis.

REMEMBERING

1 **Define** 'mitosis'.

2 **List** four types of asexual reproduction.

3 **Define** 'fertilisation'.

4 **Define** 'zygote'.

5 **Copy** the diagram of the flower in your workbook. Label its reproductive parts and state whether they are male or female.

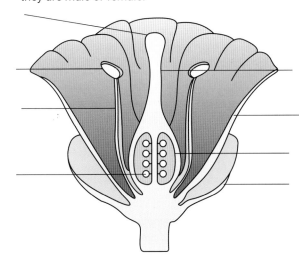

6 **Assign** the letter A for asexual and S for sexual to the following statements.

   a   Flowering plants reproduce in this way.

   b   Bacteria reproduce in this way.

   c   Offspring are genetically different from the parent.

   d   Only one parent is needed.

   e   Variation is created.

   f   Gametes are involved.

UNDERSTANDING

7 **Compare** the advantages and disadvantages of sexual and asexual reproduction.

8 **Describe** the journey of sperm from the penis to the site of fertilisation.

9 **Describe** the role of pollinators in the reproduction of flowering plants.

10 **Outline** the main functions of the cervix.

11 **Explain** the difference between budding and fragmentation.

APPLYING

12 **Relate** the structure of sperm to its function.

13 **Describe** the steps that occur during meiosis.

14 **Describe** the impact of parental care on survival rate of offspring. Use an example to justify your answer.

15 **Describe** spawning.

16 **Describe** in vitro fertilisation.

17 **Explain** why farmers prefer vegetative propagation for some of their crops. Why might this turn out to be a disadvantage?

ANALYSING

18 Parthenogenesis is classified as an asexual form of reproduction. **Explain** why it could more accurately be described as an 'incomplete form of sexual reproduction'.

19 **Discuss** the ethical issues surrounding genetic testing in IVF.

20 **Compare** the processes and animal groups involved in internal fertilisation and external fertilisation by writing two statements about the similarities and differences.

EVALUATING

21 Wasps are insects that can perform sexual and asexual reproduction. **Evaluate** this capability in terms of survival.

wikipedia/Arbeiterreserve (CC BY 4.0)

22 Endometriosis is a disorder that affects about 10 per cent of human females globally. When extra tissue grows outside the uterus, it breaks down and bleeds like a normal period. However, it is much more painful, and the tissue can get trapped. Sometimes the ovaries grow cysts or scar tissue, which prevents the release or blocks the journey of an ovum, as shown below.

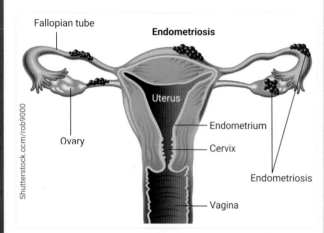

Evaluate the impact of endometriosis on pregnancy rates. **Justify** your answer.

23 Look at the following image. **Describe** what you see and relate the type of fertilisation to the number of eggs and the environment

24 **Create** a hand-drawn or digital story board showing the journey of sperm from the vagina to the fallopian tubes and ending with fertilisation.

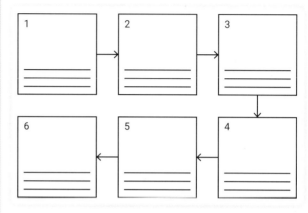

25 **Create** a poster summarising the different types of reproductive methods covered in this module: sexual, asexual, binary fission, budding, fragmentation, parthenogenesis and vegetative propagation.

26 **Make** a mind map of the methods of plant reproduction, types of pollination and methods of seed dispersal.

BIG SCIENCE CHALLENGE PROJECT #3

1 Connect what you've learned

In this chapter, you have learned about reproductive methods that help species survive by producing offspring. Are there things we can do to help our offspring, our human babies, be healthy and have the best chance of survival?

2 Check your thinking

When humans have babies, where do the babies develop?

Do babies get affected by what the pregnant mother eats, drinks and inhales?

3 Make an action plan

Prepare statements based on facts to respond to the following arguments from people who are pro-vaping.

1 E-cigarettes are not addictive.

2 E-cigarettes cannot harm babies.

3 E-cigarettes just smell good; there are no toxins.

4 Communicate

Design an infographic like the one shown here to share with your class. The purpose of the infographic is to warn them of the disadvantages of e-cigarettes.

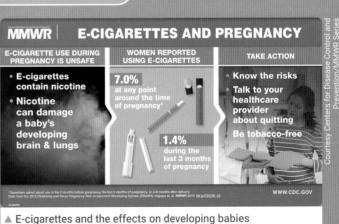

▲ E-cigarettes and the effects on developing babies

Atomic structure

4.1 **Atomic structure** (p. 104)

Atoms contain subatomic particles. The number and arrangement of these particles determines the element.

4.2 **Atomic theory** (p. 106)

The model of the atom has been developed over time by many scientists experimenting and proposing hypotheses.

4.3 **Atomic number and mass number** (p. 110)

The atomic number is the number of protons in an atom. The mass number is the number of protons and neutrons in an atom.

4.4 **Isotopes** (p. 111)

Isotopes are atoms of an element with different numbers of neutrons.

4.5 **Radioactive decay** (p. 114)

Forces in the nucleus result in unstable isotopes that emit radiation as they decay.

4.6 **Alpha, beta and gamma radiation** (p. 118)

Alpha, beta and gamma radiation are produced by the radioactive decay of radioisotopes.

4.7 **Radioactive dating** (p. 122)

The decay of carbon-14 and uranium-238/235 can be used to date materials such as animal and plant remains, and rocks.

4.8 **Uses of radiation** (p. 125)

Radioactive isotopes have a wide variety of uses in medicine and industry. Isotopes are selected for their specific properties.

4.9 **FIRST NATIONS SCIENCE CONTEXTS: First Nations Australians' history** (p. 128)

Explore dating methods that provide scientific evidence of how long First Nations Peoples have lived on the Australian continent.

4.10 **SCIENCE AS A HUMAN ENDEAVOUR: Using radioactive isotopes to track contaminants and pollutants in the environment** (p. 130)

Scientists use radioactive isotopes to identify and monitor pollution in the environment.

4.11 **SCIENCE INVESTIGATIONS: Presenting data in tables** (p. 132)

1 Simulating radioactive decay   2 Data analysis – making predictions about radioactivity

▲ **FIGURE 4.0.1** These signs warn about the dangers of radioactivity.

When most people hear the word 'radioactivity', images such as the warning sign in the photo usually spring to mind. However, it is also important to consider how we use radioactivity for our benefit. Radiation is used in medicine, to manufacture materials and even to produce some food.

▷ What do you know about radioactivity?

▷ What have TV, books, movies or comic books told you about radiation?

▷ How do the warning signs in the image above make you feel?

▷ Where do you think you encounter radiation every day?

▷ Do you know any examples of how radiation is used in medicine or industry?

#4 SCIENCE CHALLENGE ACCEPTED!

At the end of this chapter, you can complete the Big Science Challenge Project #4. You can use the information you learn in this chapter to complete the project.

Assessments
- Prior knowledge quiz
- Chapter review questions
- End-of-chapter test
- Portfolio assessment task: Science investigation and data analysis

Videos
- Science skills in a minute: Presenting data in tables (4.11)
- Videos activities: Atom structure: electron shells (4.1); Radioactive half-life (4.5); Radioactive dating (4.7); Uses of radiation (4.8); OPAL nuclear reactor (4.10)

Science skills resources
- Science skills in practice: Data tables (4.11)
- Extra science investigations: Identifying elements (4.2); Modelling isotopes (4.4)

Interactive resources
- Simulations: Rutherford scattering (4.2); Build an atom (4.3); Isotopes and atomic weight (4.4)
- Label: Parts of an atom (4.1)
- Drag and drop: Alpha, beta and gamma radiation (4.6)

❋ Nelson MindTap

To access these resources and many more, visit:
cengage.com.au/nelsonmindtap

Video activity
Atom structure: electron shells

Interactive resource
Label: Parts of an atom

atom
the fundamental particle of matter; made up of protons, neutrons and electrons

subatomic particle
a particle inside an atom, such as a proton, a neutron or an electron

proton
a positively charged particle in the nucleus of an atom

neutron
a particle in the nucleus of an atom that does not have an electrical charge

electron
a negatively charged particle in an atom, which moves in space around the nucleus

shell (electron shell)
an energy level around the nucleus of an atom containing electrons of the same energy

GET THINKING

How are atoms and elements linked? What is it about an atom that makes it oxygen, carbon or hydrogen? As you complete this module, think about ways to represent the composition of the atom in written form.

Particles in the atom

All **atoms** have a common structure and most contain the same three subatomic particles – protons, neutrons and electrons. Collectively, these are called **subatomic particles** and, when arranged in a specific way, they make up all atoms.

Protons are positively charged particles. **Neutrons** are neutral particles (have zero charge) and are about the same mass as a proton. **Electrons** are negatively charged particles and are about 2000 times smaller than a proton or a neutron. Table 4.1.1 describes the subatomic particles in an atom.

▼ **TABLE 4.1.1** Subatomic particles in an atom

Subatomic particle	Symbol	Location in the atom	Charge	Relative mass (atomic mass unit – amu)
Proton	p	Nucleus	+1	1
Neutron	n	Nucleus	0	1
Electron	e	In shells around the nucleus	−1	$\frac{1}{1840}$

Structure of the atom

All atoms have a nucleus that includes at least one proton, and all atoms have at least one electron in a **shell** (an energy level) around the nucleus. The simplest element, hydrogen, has one proton in the nucleus and one electron in a shell. The simplest form of hydrogen does not have any neutrons in the nucleus. Later in this chapter, you will learn about different forms of the same element. We call these forms isotopes.

All other elements have protons and neutrons in the nucleus and electrons around the nucleus. For example, an atom of carbon has six protons and six neutrons in the nucleus and six electrons around the nucleus in shells (Figure 4.1.1).

Atoms are electrically neutral. In all atoms, the number of positive protons is equal to the number of negative electrons. This is why an atom has no overall charge.

How are atoms of elements different?

Atoms of the same element always have the same number of protons. It is the number of protons in the nucleus that defines the element. For example, all hydrogen atoms have one proton in their nucleus, all carbon atoms have six protons, and all uranium atoms have 92 protons.

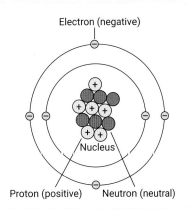

▲ **FIGURE 4.1.1** An atom of carbon showing the particle arrangement

☆ **ACTIVITY**

The relative sizes of particles in the atom

You need

- tape measure that will measure at least 2 m
- piece of chalk or a marker
- butcher's paper

What to do

1 Draw a line that is 1 mm long. This will represent an electron.

2 A proton or a neutron is approximately 2000 times larger than an electron. Calculate the length of the line you would need to draw that is 2000 times larger than the 1 mm line you just drew. Check your answer with other groups or your teacher to see if you are correct.

3 Draw your new line. This represents the size of a proton.

What do you think?

1 Did the size difference surprise you? In what way?

2 Is it easier to 'see' the difference by looking at numbers or with an activity like this? Why do you think this is?

4.1 LEARNING CHECK

1 **Define** 'element'.

2 **Describe** the mass, charge and location in the atom of protons, neutrons and electrons.

3 **Create** a model of a helium atom that shows a nucleus with two protons and two neutrons. Use materials you can easily find around your home or classroom. This could be tennis balls, ping pong balls, plasticine, even food!

　a Which features of your model are accurate and help you understand the structure of a helium atom?

　b Which features of your model could be misleading or confusing?

　c **Suggest** two improvements that would make your model a more accurate representation of a helium atom.

Interactive resource
Simulation: Rutherford scattering

Extra science investigation
Identifying elements

element
a substance consisting only of atoms that have the same number of protons in their nuclei

GET THINKING

What is the one thing that all matter in the universe has in common? Everything is made of atoms! List three things you already know about atoms and elements.

What is atomic theory?

You may recall that matter is made of elements such as oxygen, magnesium and carbon. You also may know that **elements** are made of atoms. An atom is the smallest particle of matter that keeps the chemical properties of the element.

All atoms are made up of subatomic particles – protons, neutrons and electrons.

How do we know about atoms?

About 460 BCE (nearly 2500 years ago), the Greek philosopher Democritus was the first person to propose the concept of an atom. His theory suggested that if you kept cutting something into smaller pieces, you would eventually get a piece that could not be cut any smaller (Figure 4.2.1). The word 'atom' comes from the Greek word *atomos*, which means 'indivisible' (cannot be divided). Since then, many scientists have performed experiments that confirmed the existence of atoms. Scientists also discovered that within atoms are even smaller subatomic particles, and they worked out how the subatomic particles are arranged in an atom.

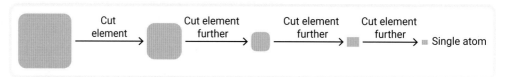

Cut element → Cut element further → Cut element further → Cut element further → ■ Single atom

▲ **FIGURE 4.2.1** If it was possible to cut an element into smaller and smaller pieces, you would eventually reach the smallest form of that element – the atom.

As scientific knowledge increased, scientists were able to perform experiments that confirmed or rejected theories about atoms. The problem with atoms is that, even with modern technology such as electron microscopes, we cannot see them to confirm their structure. Thus, all descriptions of the atom we have are models.

Models are useful because they help us to visualise something too small to see. Models are not perfect and are improved over time as discoveries are made and new technology is invented. There have been many versions of the atomic model and it is still being developed.

Dalton's model of the atom

John Dalton (1766–1844) proposed the first modern model of the atom. He performed experiments and examined the results of other scientists' experiments. His conclusion was the same as that of Democritus. The results could only be explained if matter was made of small particles. He made several important conclusions about atoms and elements that are still accepted today (Figure 4.2.2).

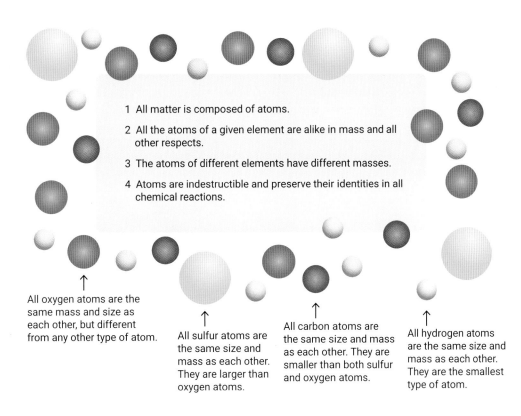

1 All matter is composed of atoms.

2 All the atoms of a given element are alike in mass and all other respects.

3 The atoms of different elements have different masses.

4 Atoms are indestructible and preserve their identities in all chemical reactions.

All oxygen atoms are the same mass and size as each other, but different from any other type of atom.

All sulfur atoms are the same size and mass as each other. They are larger than oxygen atoms.

All carbon atoms are the same size and mass as each other. They are smaller than both sulfur and oxygen atoms.

All hydrogen atoms are the same size and mass as each other. They are the smallest type of atom.

▲ **FIGURE 4.2.2**   Dalton's conclusions about the atom

Thomson's plum pudding model

Since the 1850s, scientists have conducted experiments with **cathode ray** tubes.

- Cathode ray tubes pass an electric current through a piece of metal called a cathode.

- In the 1800s, scientists produced a glowing beam (see Figure 4.2.3), but no-one knew what it was made from.

- When the metal cathode was changed to a different metal element, an identical glowing beam was produced.

- Scientists concluded that all metals (and all materials used) made exactly the same beam.

- This meant that something in atoms was producing the beam. But they did not know what it was!

cathode ray
a beam of electrons produced by a cathode ray tube

▲ **FIGURE 4.2.3**   A cathode ray tube fires electrons that can be bent by electric and magnetic fields.

The British scientist J.J. Thomson (1856–1940) performed experiments in 1897 that showed the beam could be bent by an electric field. He concluded that the beam was made of tiny negatively charged particles. He knew they were negative because they were attracted to a positive charge inside the cathode ray tube. We now know these negative particles as electrons. Every material used in the cathode ray tube produced the same particles. Thus, Thomson concluded that the particles (electrons) were found in all types of atoms.

Thomson proposed a model of the atom that advanced the previous Dalton model. Atoms are usually electrically neutral, so the fact that they contained negative particles meant they also had to contain the same number of positive particles. Thomson proposed that the negatively charged electrons were contained inside a positively charged sphere – like fruit spread through a Christmas pudding or chocolate chips in a muffin. This is why Thomson's model of the atom is often referred to as the 'plum pudding model' (Figure 4.2.4).

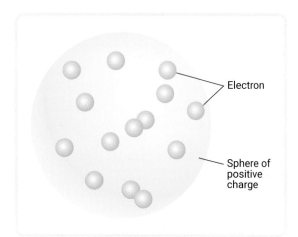

Electron

Sphere of positive charge

▲ **FIGURE 4.2.4** Thomson's plum pudding model. A muffin with chocolate chips is a good model for the plum pudding model. In this model, the muffin is the positive charge and the chocolate chips are the electrons.

Rutherford

New Zealand scientist Ernest Rutherford (1871–1937) further investigated the structure of the atom by testing Thomson's model. Rutherford used very thin gold foil, just a few atoms thick, and fired **alpha particles** at the foil (Figure 4.2.5). An alpha particle is a helium nucleus, consisting of two protons and two neutrons, which gives it an overall positive charge. If the gold atom were like a plum pudding, then all the alpha particles should pass through the foil with minimal change of direction.

Rutherford was surprised to find that a small proportion of the alpha particles were deflected or bounced backwards. There was something in the gold atoms that blocked the path of the alpha particles. Rutherford suggested this was due to the atoms having a positive charge, concentrated in a structure he called the nucleus. This positive nucleus was repelling the positive alpha particles and changing their path. Therefore, he concluded that the plum pudding model was incorrect.

alpha particle
a helium nucleus (two protons and two neutrons) emitted when unstable larger nuclei decay

9780170472838

Rutherford proposed a new model of the atom. He proposed that atoms:

- were mostly empty space
- had positively charged protons in a nucleus in the centre
- had negatively charged electrons outside the nucleus.

Bohr

One of the problems with Rutherford's model was that he did not know how the electrons were arranged. He proposed they orbited like planets around the nucleus. Niels Bohr (1885–1962) proposed that electrons were arranged in shells, or energy levels, around the nucleus (Figure 4.2.6). His theory was that electrons in the same shell had the same energy. Bohr also suggested that if energy was added to the atom, the electrons could move up to higher shells, then emit energy when they returned to the original shell. This theory explained some features of atoms that scientists had not been able to explain before.

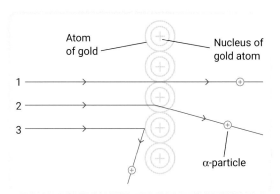

▲ **FIGURE 4.2.5**   Rutherford's gold foil experiment

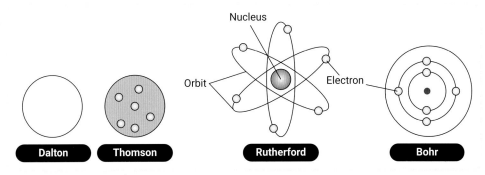

▲ FIGURE 4.2.6   The development of the model of the atom

Chadwick's discovery of the neutron

The model of the atom still had some problems that scientists couldn't explain, with the mass of the nucleus not matching the proposed particles and structure. The suggested arrangement of protons and electrons had a smaller mass than the actual mass measured. The atom was too heavy, indicating some other particles were present. James Chadwick proposed that the nucleus also contained electrically neutral particles. He conducted experiments to show that this particle was present in atoms. We know this particle as the neutron.

4.2 LEARNING CHECK

1 **State** three features of atoms as proposed by Dalton.
2 **Describe** the differences between Dalton's and Thomson's models of the atom.
3 **Describe** Rutherford's experiment and how it advanced the previous atomic model.
4 **Explain** one benefit and one limitation of atomic models.
5 **Create** a series of labelled diagrams to show how the atomic model has changed over time. Include the name of the scientist(s) and part(s) of the atom discovered.

Interactive resource
Simulation: Build an atom

GET THINKING

Using numbers and chemical symbols gives important information about the subatomic parts of an atom. When you see scientific notation like C-12 or U-238, what questions do you have?

Determining the number of particles in an atom

Scientists have developed a convention for writing about the number of subatomic particles in an atom. Scientific conventions like these are important so that scientists and students across the world interpret and present chemical information in the same way.

As you learned in the previous module, atoms of the same element have the same number of protons in the nucleus. This number is also known as the **atomic number** of an element. The atomic number is given the symbol Z. Because the number of protons and electrons in an atom is balanced, this number also tells us about the number of electrons in a nucleus.

atomic number
the number of protons in the nucleus of an atom, which is the same for every atom of the same element; symbol Z

mass number
the total number of protons and neutrons in the nucleus of an atom; symbol A

The **mass number** is the total number of particles in the nucleus; that is, the number of protons and neutrons added together. The mass number is given the symbol A.

Mass number = number of protons + number of neutrons

Elements are often referred to by their mass number together with their name or symbol. Carbon-12 or C-12 is a carbon atom with a mass number of 12. Uranium-235 shows that uranium has a mass number of 235.

Mass number ⟶ 12
Atomic number ⟶ 6 C

▲ **FIGURE 4.3.1** Notation showing mass number and atomic number for carbon-12

The atomic number and mass number can be placed next to the element symbol as shown in Figure 4.3.1.

You can use this notation to determine the number of protons, neutrons and electrons. For example, the notation for chlorine-35 or Cl-35 is $^{35}_{17}\text{Cl}$. From this information you can work out that an atom of chlorine has:

- 17 protons (from the atomic number of 17)
- 17 electrons (same number of protons and electrons)
- 18 neutrons (35 − 17 = 18, or mass number − atomic number = number of neutrons).

4.3 **LEARNING CHECK**

1 What information is provided by the atomic number and the mass number of an element and how are they different?

2 **Explain** why you cannot get the full information about atomic structure from the term 'oxygen-16' without any other information.

3 **Identify** the number of protons, neutrons and electrons in the following atoms.
   a $^{56}_{26}\text{Fe}$ b $^{235}_{92}\text{U}$

Isotopes

BY THE END OF THIS MODULE, YOU WILL BE ABLE TO:

✓ compare different isotopes of an element

✓ create a model to show the isotopes of a specific element.

GET THINKING

Look at Figure 4.4.1. From what you learned in the previous module, why do you think it is possible to have carbon-12, carbon-13 and carbon-14?

What is an isotope?

Three **isotopes** of carbon are shown in Figure 4.4.1. They are all referred to as 'isotopes of carbon'.

isotopes
atoms of an element with the same number of protons but different numbers of neutrons

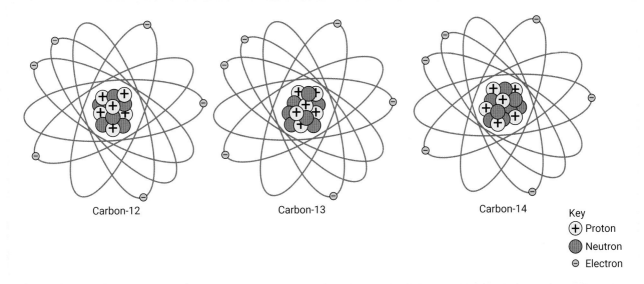

Carbon-12   Carbon-13   Carbon-14

Key
(+) Proton
⬤ Neutron
⊖ Electron

▲ **FIGURE 4.4.1**   The three isotopes of carbon

If you look carefully at the diagrams, you can see some similarities and differences between the three carbon isotopes.

Similarities:

- same number of protons in the nucleus (same atomic number)
- same number of electrons around the nucleus.

Differences:

- different number of neutrons in the nucleus.

Interactive resource
Simulation: Isotopes and atomic weight

Extra science investigation
Modelling isotopes

▼ **TABLE 4.4.1**   Similarities and differences of carbon isotopes

Element	Atomic number (Z)	Number of protons	Number of neutrons	Mass number (A)
Carbon-12	6	6	6	12
Carbon-13	6	6	7	13
Carbon-14	6	6	8	14

Isotopes are atoms of an element with the same number of protons but different numbers of neutrons. Therefore, isotopes have a different mass number because of the different number of neutrons.

Which elements have isotopes?

Many elements have multiple isotopes that can be either natural or artificial (Figure 4.4.2).

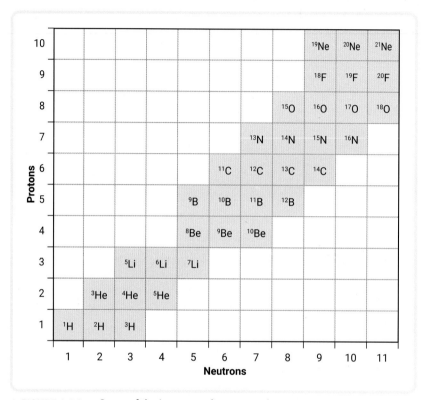

▲ **FIGURE 4.4.2** Some of the isotopes of common elements

natural isotope
an isotope of an element found in nature

Elements with an atomic number of 93 or greater have no **natural isotopes** and are produced in laboratories. These can be seen in pink on the periodic table in Figure 4.4.3. Some elements with atomic numbers less than 93 have isotopes that do not exist in nature but can also be made artificially.

artificial isotope
an isotope of an element produced in a nuclear reactor or particle accelerator

One way these **artificial isotopes** can be produced is in nuclear reactors by adding particles to the nucleus of large elements such as uranium and thorium. This process is known as bombardment (collision) and can occur with protons or neutrons. For example, plutonium-239 can be converted to americium-240 by adding a neutron through neutron bombardment.

particle accelerator
a machine that enables high-speed, high-energy collisions between atoms to produce artificial isotopes and new elements

Artificial elements can also be produced in **particle accelerators** by high-speed, high-energy collisions of particles.

	1												18

Periodic table

| | **1** | | **2** | | **13** | **14** | **15** | **16** | **17** | **18** |

Row 1: 1 H (group 1), 2 (group 2); 2 He (group 18)

Row 2: 3 Li, 4 Be; 5 B, 6 C, 7 N, 8 O, 9 F, 10 Ne

Row 3: 11 Na, 12 Mg; groups 3–12; 13 Al, 14 Si, 15 P, 16 S, 17 Cl, 18 Ar

Row 4: 19 K, 20 Ca, 21 Sc, 22 Ti, 23 V, 24 Cr, 25 Mn, 26 Fe, 27 Co, 28 Ni, 29 Cu, 30 Zn, 31 Ga, 32 Ge, 33 As, 34 Se, 35 Br, 36 Kr

Row 5: 37 Rb, 38 Sr, 39 Y, 40 Zr, 41 Nb, 42 Mo, 43 Tc, 44 Ru, 45 Rh, 46 Pd, 47 Ag, 48 Cd, 49 In, 50 Sn, 51 Sb, 52 Te, 53 I, 54 Xe

Row 6: 55 Cs, 56 Ba, *, 72 Hf, 73 Ta, 74 W, 75 Re, 76 Os, 77 Ir, 78 Pt, 79 Au, 80 Hg, 81 Tl, 82 Pb, 83 Bi, 84 Po, 85 At, 86 Rn

Row 7: 87 Fr, 88 Ra, **, 104 Rf, 105 Db, 106 Sg, 107 Bh, 108 Hs, 109 Mt, 110 Ds, 111 Rg, 112 Cn, 113 Nh, 114 Fl, 115 Mc, 116 Lv, 117 Ts, 118 Og

* Lanthanoids: 57 La, 58 Ce, 59 Pr, 60 Nd, 61 Pm, 62 Sm, 63 Eu, 64 Gd, 65 Tb, 66 Dy, 67 Ho, 68 Er, 69 Tm, 70 Yb, 71 Lu

** Actinoids: 89 Ac, 90 Th, 91 Pa, 92 U, 93 Np, 94 Pu, 95 Am, 96 Cm, 97 Bk, 98 Cf, 99 Es, 100 Fm, 101 Md, 102 No, 103 Lr

▲ **FIGURE 4.4.3** The periodic table with elements with no naturally occurring isotopes shown in pink (current as of October 2022)

4.4 LEARNING CHECK

1 Lithium-5, lithium-6 and lithium-7 are three isotopes of lithium. **Identify** one thing the isotopes have in common, and one way they are different.

2 **Describe** the difference between natural and artificial isotopes.

3 Helium-3, helium-4 and helium-5 all have two protons in the nucleus. Draw diagrams or create a physical model (using plasticine or coloured balls) to show the arrangement of protons, neutrons and electrons in the three helium isotopes.

4 Use an example to **explain** one way that artificial isotopes are produced.

5 The images A–F show atoms that may or may not be isotopes of each other.

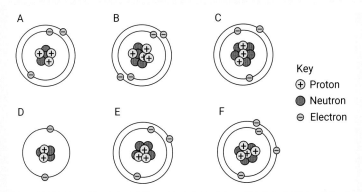

Key
⊕ Proton
● Neutron
⊖ Electron

a **Identify** three atoms that are isotopes of each other. Justify your selection.

b **Identify** the atom that does not have any isotopes. Justify your selection.

c Are there any other isotopes (different from in parts **a** and **b**) in the image? **Explain** your answer.

Video activity
Radioactive half-life

radioactive isotope (radioisotope)
an isotope that is unstable and undergoes radioactive decay to become more stable

radiation
a stream of particles and/or energy from a radioactive source

electrostatic force
a force of attraction between oppositely charged particles

strong nuclear force
a force of attraction between particles in the nucleus of an atom

GET THINKING

'Radioactive' is a word that is misunderstood by a lot of people. Write down what you would think if someone told you there were radioactive elements in the same room as you!

Stable and unstable isotopes

You may recall from previous modules that carbon has a number of isotopes. Some of these isotopes, such as carbon-12 and carbon-13, are stable. Other isotopes, such as carbon-14, are unstable. Unstable isotopes are known as **radioactive isotopes** and release **radiation**.

How do you get stable and unstable isotopes?

Inside all nuclei two opposing forces are acting. You may recall that when you have two identical charges (negative and negative or positive and positive), they will experience a repulsive force. This **electrostatic force** acts between the protons in the nucleus as they push each other away because of their identical, positive charges. There is also a force called the **strong nuclear force** that attracts and holds the protons and neutrons together in the nucleus.

When these two forces are balanced, the isotope is stable. This happens when the attractive and repulsive forces are about the same strength. When one of the forces is much stronger than the other force, the unbalanced forces cause the isotope to be unstable.

Proton/neutron balance

The strong nuclear force is attractive and acts between all particles in the nucleus. It acts between protons and neutrons, which have no charge. This may seem strange because like charges usually repel. However, that is the *electrostatic* force. The *strong nuclear* force is another type of force that works in a different way. It is attractive between all particles in close contact, regardless of charge. It only operates in the nucleus of atoms (Figure 4.5.1).

Therefore, to make a nucleus stable, you need neutrons to provide stronger nuclear force and oppose the repulsive force between protons. This is why stable isotopes have a certain number of protons and neutrons that allow the nucleus to be stable.

For smaller elements, such as the first 20 elements, the stable ratio is approximately 1 proton to 1 neutron. For larger elements, more neutrons are required to balance the number of protons, so you see the ratio change.

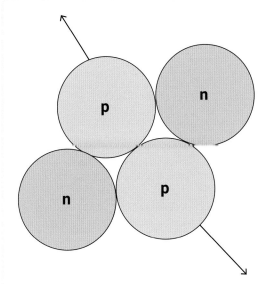

Electrostatic forces repel the protons, pushing them away from each other.

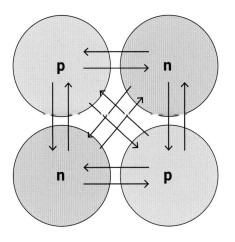

The strong nuclear force holds all particles in the nucleus together.

▲ **FIGURE 4.5.1** Isotopes are stable when electrostatic and nuclear forces are balanced.

Lead is the largest stable element with 82 protons (atomic number 82). Atoms larger than lead do not have enough neutrons to balance the forces, so will never be stable. This means elements with an atomic number greater than 82 have no stable isotopes. Table 4.5.1 shows some stable isotope proton-to-neutron ratios.

What is radioactive decay?

Unstable isotopes cannot stay in their unstable form and will release particles or energy from their nucleus to become stable isotopes. Because the reason for instability is an unbalanced proton-to-neutron ratio, there must be changes to this ratio to make the nucleus stable again. In the next module, we will look at three common processes that occur in unstable isotopes.

▼ **TABLE 4.5.1** Some stable isotope proton-to-neutron ratios. As the atom gets larger, the ratio increases

Isotope	Protons	Neutrons	Proton-to-neutron ratio
Helium-4	2	2	1:1
Carbon-12	6	6	1:1
Calcium-40	20	20	1:1
Scandium-44	21	23	1:1.1
Silver-107	47	60	1:1.3
Lead-206	82	124	1:1.5

The process of changing the particles in the nucleus often results in the production of particles and/or the emission of energy from the isotope. This nuclear change is spontaneous (happens without any external factor) and is called **radioactive decay**. Isotopes that undergo radioactive decay are called radioactive isotopes, or radioisotopes.

Half-life

When a radioactive isotope decays, it does so in a predictable way. All radioisotopes have a certain **half-life**. The half-life of an isotope is the time it takes for half the nuclei to decay.

Half-lives vary significantly. Some radioactive isotopes exist for only fractions of seconds. Fermium-244 has a half-life of 3.3 milliseconds. Others can exist for minutes, hours or years. Radon-222 has a half-life of 3.82 days and uranium-238 has a very long half-life of 4.5 billion years.

Calculating half-life

The half-life of cobalt-60 is 5.27 years. This means that if you start with 10 grams of cobalt-60, there would only be 5 grams (half) left after 5.27 years. As seen in Figure 4.5.2, every 5.27 years another half of the sample has decayed. So, after 10.54 years (2 half-lives) only 2.5 grams will remain.

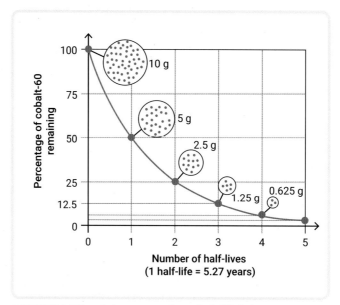

▲ FIGURE 4.5.2   A decay curve for cobalt-60. A decay curve can be plotted with the number of half-lives or time on the horizontal axis.

If you have a decay curve, you can determine the half-life of a radioactive isotope, or find out how long it takes to decay to a certain level. Look at Figure 4.5.3. You could use this decay curve in a number of ways.

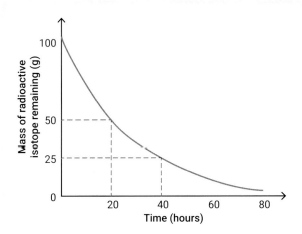

▲ **FIGURE 4.5.3** A decay curve for a radioisotope

1 Calculating the half-life by finding half the original mass (50 grams) and using the graph to find out the time this takes:

* rule a horizontal line from mass = 50 grams on the vertical axis to the curve
* rule a horizontal line down to the horizontal axis to determine the time, this gives a value of 20 hours.

2 Determining how much mass would be left after a period of time. If you wanted to know how much of the sample was left after two half-lives, you could use the half-life of 20 hours to calculate this. Two half-lives are 40 hours. Rule lines (as described above) but this time from 40 hours on the horizontal axis up to the curve, then across to the vertical axis to find the mass remaining. In this case, 25 grams of the sample remain.

You can also calculate half-life information from simple data. For example, sodium-20 has a half-life of 15 hours. If you start with 100 g, how much will remain after 45 hours?

* $\dfrac{45 \text{ hours}}{15 \text{ hours}} = 3$ half-lives
* One half-life reduces the sample from 100 g to 50 g.
* The second half-life reduces the sample to 25 g.
* The third half-life reduces the sample to 12.5 g.
* Thus, after 45 hours (or 3 half-lives), 12.5 g of the sample will remain.

4.5 LEARNING CHECK

1 **Define** 'radioactive isotope'.
2 In terms of the strong nuclear and electrostatic forces in the nucleus, **explain** what makes an isotope unstable.
3 If an atom of magnesium had 12 protons and 15 neutrons, **explain** whether it would be stable or unstable.
4 A sample of the radioactive isotope caesium-134 has a half-life of 2.06 years. If you started with a 200 g sample, how long would it take for 6.25 g of the sample to remain?
5 Use examples to show how the proton-to-neutron ratio increases as the atomic mass increases.

Alpha, beta and gamma radiation

Interactive resource
Drag and drop: Alpha, beta and gamma radiation

GET THINKING

Look at the diagrams of the types of nuclear decay in this module (Figures 4.6.1–4.6.3). What similarities and differences do you see? List at least two similarities and two differences.

Radioactive decay

In the previous module, you learned that the proton-to-neutron ratio of a nucleus determines the stability of the isotope. When the proton-to-neutron ratio is either too high or too low, the isotope is radioactive and will decay.

When a radioactive isotope decays, its proton-to-neutron ratio becomes closer to that of a stable isotope. Not all radioactive decay ends with a fully stable isotope, but the result will always be a more stable isotope than the starting isotope.

Three main types of radiation are emitted during radioactive decay – alpha particles, beta particles and gamma radiation.

Alpha particle production

Alpha particles are primarily produced when very large atoms decay. You may recall that no element of atomic number greater than 82 can be stable. To become more stable, these large radioisotopes lose protons and neutrons to reduce the mass of the nucleus. The most energy-efficient method of doing this is for the nucleus to emit an alpha particle. Alpha particles consist of two protons and two neutrons; they are helium nuclei (helium atoms without any electrons).

Emitting a particle with multiple protons and neutrons means the isotope (parent nucleus) loses a lot of nuclear mass quickly. The resulting product (daughter nucleus) is more stable than the initial isotope because its nucleus is smaller.

An example of alpha particle production is the decay of radon-222 to polonium-218 (Figure 4.6.1 and Table 4.6.1).

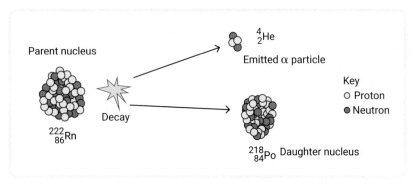

▼ **TABLE 4.6.1** The alpha decay of radon-222 to polonium-218

Isotope	Atomic number	Mass number
Radon-222	86	222
Polonium-218	84	218

▲ **FIGURE 4.6.1** The alpha decay of radon-222 to polonium-218

Beta particle production

Beta particles are negatively charged particles emitted from an unstable atom's nucleus during radioactive decay. The resulting nucleus has a more stable proton-to-neutron ratio than the original atom. Beta radiation can come from unstable nuclei with an excess of neutrons or protons.

beta particle
a negatively charged particle identical to an electron, emitted when an unstable nucleus decays

An example of beta decay is the conversion of carbon-14 into nitrogen-14 with the emission of a beta particle (Figure 4.6.2).

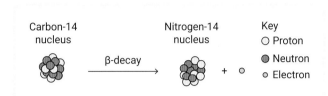

▲ **FIGURE 4.6.2** The beta decay of carbon-14 to nitrogen-14

Gamma radiation production

Gamma radiation is high-energy radiation emitted during radioactive decay. Gamma radiation does not involve particles. It is just energy. An example of gamma radiation production involves cobalt-60. This isotope decays to produce beta particles. The resulting high-energy nickel-60 daughter nuclei emits this extra energy as gamma radiation, forming a low-energy nickel-60 nucleus (Figure 4.6.3).

gamma radiation
high-energy electromagnetic energy emitted when an unstable nucleus decays

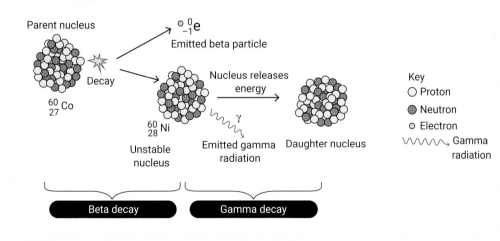

▲ **FIGURE 4.6.3** The beta and gamma decay of cobalt-60

Properties of alpha, beta and gamma radiation

The three types of radiation have different properties, which allow them to be used for different purposes.

One property of all radiation is **penetrating power**. This describes its ability to pass through materials (Figure 4.6.4). Alpha particles are large and highly charged so they don't pass through materials easily. Beta particles are smaller in mass and have a smaller charge. They can pass through materials more easily than alpha particles. Gamma radiation has no mass and no charge because it is electromagnetic radiation and can pass through some dense materials.

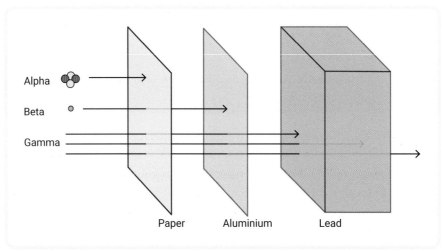

▲ **FIGURE 4.6.4** Alpha particles are stopped by paper, beta particles can be blocked by aluminium, whereas gamma radiation needs very thick lead or concrete to block it.

Radiation also has the property of **ionising power**. Radiation can change the structure of materials it passes through by causing atoms to lose electrons and form positively charged particles. Particles like this with a positive or negative charge are called ions. Ionisation of atoms changes the way they function. When this happens to the DNA or proteins in the tissue of living things, it can cause cancer, cell death or other cell damage. Alpha particles are the strongest ionising radiation because their mass and charge are more likely to affect the electrons in atoms. Beta particles with lower mass and charge are less likely to affect atoms, while gamma radiation usually causes damage due to energy transfer inside the material.

The true damage that radiation causes is a combination of both ionising power and penetrating power. Alpha particles cause a lot of damage but cannot pass through skin. Alpha radiation is only dangerous to humans if it is accidentally ingested (swallowed). Gamma radiation is the type of radiation most likely to pass into human tissues and cause internal damage, as seen in Figure 4.6.5.

9780170472838

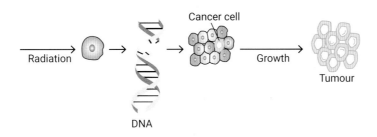

FIGURE 4.6.5 DNA damage due to ionising radiation

4.6 LEARNING CHECK

1 **Describe** the composition of alpha particles, beta particles and gamma radiation in words, or with labelled diagrams.

2 **Create** a table to **summarise** and **compare** the relative penetrating power and ionising power of the different types of radiation. Use 'high', 'medium' and 'low' to rank the three types of radiation.

3 **Predict** the atomic number and mass number of the daughter nucleus that forms when $^{238}_{92}U$ undergoes alpha decay.

4 **Explain** why a neutron would convert to a proton during beta decay by considering the stability of the final product.

5 A high-energy barium-137 isotope undergoes gamma decay. **Predict** and **explain** the mass number of the final daughter nuclei.

6 A scientist was trying to determine the thickness of lead that was needed to completely block the beta radiation from a sample of sodium-22. To do this, the scientist measured the radiation that passed through different thicknesses of lead. The results are shown in Table 4.6.2.

▼ **TABLE 4.6.2** Radiation that passes through different thicknesses of lead

Lead thickness (mm)	Radiation count
0	35 500
2.5	23 500
5	15 000
7.5	10 000
10	6500
12.5	4000

a **Construct** a graph with radiation count on the vertical axis, and lead thickness on the horizontal axis. The horizontal axis should cover values from 0 to 30 mm.

b **Draw** a curve of best fit and use this to **estimate** the thickness of lead that would be needed to completely block the radiation from the sodium-22 radioisotope.

c Are you confident in the value you estimated? Why or why not?

d **Suggest** an improvement to the experiment that would increase your confidence in your estimated value.

4.7 Radioactive dating

BY THE END OF THIS MODULE, YOU WILL BE ABLE TO:
✓ explain how radioactive isotopes are used to date materials
✓ calculate the age of materials using known data.

Video activity
Radioactive dating

carbon-14 dating
the use of the carbon-14 decay process to determine the age of materials containing carbon

uranium–lead dating
the use of either the uranium-238 or uranium-235 decay process to determine the age of rocks

GET THINKING

Examine the photographs in Figure 4.7.1. What element do they have in common? How do you think radioactive decay could be used to find out how old an object is?

Dating materials

One use of radioactive isotopes is for dating materials. **Carbon-14 dating** is used to estimate the age of materials that contain carbon. This includes anything that was once a living organism. **Uranium–lead dating** is used to estimate the age of rocks that are between 1 million and 4.5 billion years old. Both methods use the known half-lives of isotopes to determine how many half-lives have passed. This allows scientists to calculate the age of the material.

Carbon-14 dating

Carbon-14 dating is widely used by scientists to date material that contains carbon. This includes any plant or animal remains. Materials such as wooden buildings and ships, paper and parchments, charcoal, twigs and seeds can be carbon-dated. Animal remains including bones, shells, leather and hair can also be carbon-dated (Figure 4.7.1).

▲ **FIGURE 4.7.1** Carbon-dating can be used to estimate the age of wooden ships and human/animal remains such as those of woolly mammoths.

Carbon-14 dating uses the radioactive isotope carbon-14 (6 protons, 8 neutrons). The amount of carbon-14 on Earth is small but very consistent. Only one out of 1 000 000 000 000 atoms of carbon on Earth are carbon-14. All living things contain a fixed percentage of carbon-14. This allows scientists to calculate how much carbon-14 has decayed since the organism died.

9780170472838

Plants take in carbon dioxide from the atmosphere when they photosynthesise. A small, consistent percentage of the carbon taken in is carbon-14. This carbon-14 moves through the food chain, so the same percentage of carbon-14 is present in all living organisms. While a plant or an animal is alive, as the radioactive isotope decays, it is replaced by more carbon-14 as the plant photosynthesises or the animal eats. Therefore, if you know the mass of the plant or animal when it was alive, you can accurately determine the mass of carbon-14 it contained.

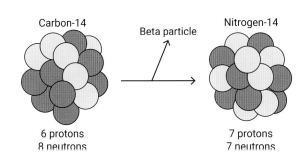

▲ FIGURE 4.7.2 The radioactive decay of carbon-14 into nitrogen-14 with release of beta particles. The beta particles are lost from the carbon-14 isotopes.

Once a plant or an animal dies, it no longer takes in new carbon-14. The carbon-14 left behind decays at a consistent rate to nitrogen-14, as seen in Figure 4.7.2.

Scientists measure the mass of carbon-14 in the remains of the plant or animal, calculating how much must have been present originally, as shown in Figure 4.7.3. The missing mass is due to decay and can be used to determine the age as follows.

- The item being tested was once alive, so a specific amount of carbon-14 was present in the object when it was a living organism.
- Scientists determine the remaining amount of carbon-14 in a sample, based on the rate of the production of beta particles as the carbon-14 decays.
- Imagine 25 per cent of the object's initial amount of carbon-14 remains. Because a half-life of an isotope is the time it takes for half its nuclei to decay, we know that it has taken two half-lives for the original amount of carbon-14 to become 25 per cent: 100 per cent (original amount) → 50 per cent (1 half-life) → 25 per cent (2 half-lives).

Because the half-life of carbon-14 is known to be 5730 years, the object is approximately 11 460 years old.

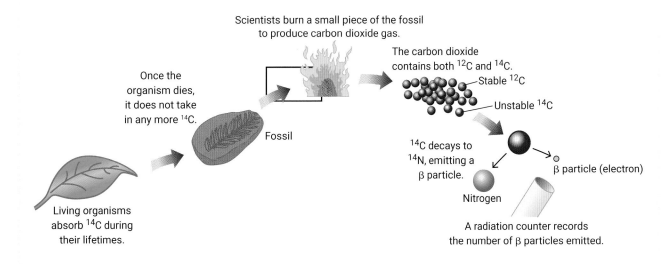

▲ FIGURE 4.7.3 Carbon-14 is present in a fixed percentage in all living objects. This is used to determine the age of the material by measuring how much has decayed.

<table>
<tr><td>

Uranium-238 decay series

$^{238}_{92}U$
↓ alpha
$^{234}_{90}Th$
↓ beta
$^{234}_{91}Pa$
↓ beta
$^{234}_{92}U$
↓ alpha
$^{230}_{90}Th$
↓ alpha
$^{226}_{88}Ra$
↓ alpha
$^{222}_{86}Rn$
↓ alpha
$^{218}_{84}Po$
↓ alpha
$^{214}_{82}Pb$
↓ beta
$^{214}_{83}Bi$
↓ beta
$^{214}_{84}Po$
↓ alpha
$^{210}_{82}Pb$
↓ beta
$^{210}_{83}Bi$
↓ beta
$^{210}_{84}Po$
↓ alpha
$^{206}_{82}Pb$

</td><td>

Uranium-235 decay series

$^{235}_{92}U$
↓ alpha
$^{231}_{90}Th$
↓ beta
$^{231}_{91}Pa$
↓ alpha
$^{227}_{89}Ac$
↓ beta
$^{227}_{90}Th$
↓ alpha
$^{223}_{88}Ra$
↓ alpha
$^{219}_{86}Rn$
↓ alpha
$^{215}_{84}Po$
↓ alpha
$^{211}_{82}Pb$
↓ beta
$^{211}_{83}Bi$
↓ alpha
$^{207}_{81}Ti$
↓ beta
$^{207}_{82}Pb$

</td></tr>
</table>

▲ **FIGURE 4.7.4** The uranium to lead decay series

Uranium–lead dating

Uranium-238 decays in a series of steps to become lead-206. Uranium-235 decays to lead-207 (Figure 4.7.4). Both isotopes of uranium are used to estimate the age of rocks. This type of dating is called uranium–lead dating.

As a rock forms, it may develop crystals. When rock crystals form, uranium atoms are trapped inside them. Over time, the uranium-238 atoms decay to lead-206, with a half-life of 4.47 billion years. The uranium-235 atoms decay to lead-207, with a half-life of 710 million years.

Both decay series can be used to date rocks by determining the percentage of uranium and lead isotopes present in the rock. In general terms, the more lead that is present, the older the rock is.

- If a rock contains 50 per cent uranium-238 and 50 per cent lead-206, this means half the uranium-238 has decayed (one half-life), and the rock is approximately 4.47 billion years old.

- If a rock contains 25 per cent uranium-235 and 75 per cent lead-207, this means the uranium-235 has been through two half-lives (100 per cent → 50 per cent → 25 per cent) and the rock is approximately 1420 million years old.

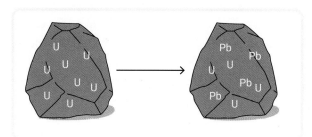

▲ **FIGURE 4.7.5** Uranium decays to lead in rocks.

4.7 LEARNING CHECK

1 **Describe** uses of carbon-14 and uranium–lead dating.

2 **Explain** why carbon-14 could not be used to determine the age of wooden buildings in Australia from the 1900s.

3 A sample of preserved wood is determined as being 28 650 years old. What percentage of carbon-14 would remain in the preserved wood now?

4 Earth is 4.54 billion years old, and rock formation is a continuous process, with rocks constantly forming throughout Earth's history. **Explain** why both uranium-238 and uranium-235 are needed to determine the age of rocks on Earth.

5 Figure 4.7.1 shows two examples of discoveries that have been made using carbon-14 dating. Choose one of the images and **research** a notable discovery in this area. **Create** a two-slide presentation showing key features of the discovery and the role of carbon-14 dating.

9780170472838

4.8 Uses of radiation

BY THE END OF THIS MODULE, YOU WILL BE ABLE TO:

✓ describe how radiation is used for different purposes

✓ explain why different types of radiation are used for different purposes

✓ predict the best type of radiation for a particular purpose.

GET THINKING

Skim over the examples of radioactive isotopes in this module. How many of them have you heard of? Make a list of two or three examples that interest you for further research later in the module.

Video activity
Uses of radiation

Uses of radioactive isotopes

Radioactive isotopes are used for a range of applications such as medicine, plumbing, construction and scientific research. The choice of radioactive isotope depends on its properties. You may recall that radiation can have high, medium or low ionising power and penetrating power. Ionising power is the ability to damage atoms. Penetrating power is the ability to pass through materials.

Radioactivity in medicine

Radioactivity has many uses in medicine. Doctors use it to find out what illness a patient is suffering from and to treat illnesses such as cancer.

Nuclear medicine monitors what happens to certain chemicals as they pass through the body. It can be used to check if an organ is functioning properly. These chemicals, called tracers, are made with radioactive isotopes (Figure 4.8.1). This allows doctors to follow their path through the body by monitoring the radiation they emit.

Isotopes used in nuclear medicine have the following properties.

- Short half-life – radioactive substances should not stay in the human body longer than necessary. Most radioactive isotopes used in medicine have a half-life of only hours or days.

- Emission of detectable radiation – the radioactive isotope should emit radiation that has a high enough penetrating power to leave the human body and be detected, but preferably a low ionising power so that it does not damage cells.

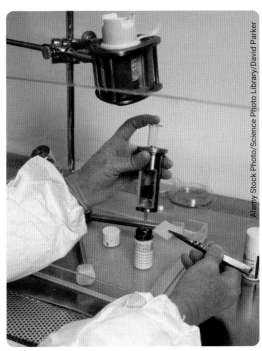

▲ **FIGURE 4.8.1** A chemical tracer containing a radioactive isotope

Alamy Stock Photo/Science Photo Library/David Parker

Isotope	Use
Technetium-99	Diagnosis of heart, lung, brain, thyroid and blood flow problems
Fluorine-18	Study of brain function (Figure 4.8.2) Diagnosis of epilepsy, heart disease and some types of cancer
Iodine-123	Diagnosis of thyroid disease and some types of cancer

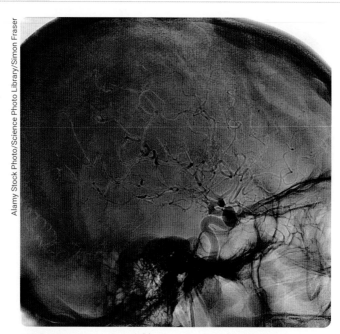

▲ **FIGURE 4.8.2** Arterial brain scans involve injecting a radioactive isotope such as fluorine-18 into a blood vessel and taking an image.

radiotherapy
the treatment of cancer by radiation

The use of radiation in the treatment of cancer is called **radiotherapy**. Cancer cells actively divide and are more likely to be killed by radiation than healthy cells. Radiotherapists target the radiation at the cancer cells.

Radiotherapy machines often use cobalt-60 as a source of gamma radiation. Gamma radiation has very high energy and if targeted correctly will damage cancer cells. It does this by transferring this energy to the chemical bonds, which break and cause cell damage and death. Cobalt-60 has a half-life of 5.2 years. This longer half-life is not a problem for the patient because the isotope is in the machine and not in the body.

▼ **TABLE 4.8.2** Medical treatment isotopes

Isotope	Half-life	Use
Iodine-131	8.03 days	Treatment of thyroid disease
Phosphorus-32	14.26 days	Treatment of excess red blood cells
Yttrium-90	64 hours	Liver cancer therapy
Iridium-132	73.83 days	Treatment of cancer in the body

9780170472838

Radioactivity in industry

Radioactive isotopes have industrial applications. Caesium-137 is used to determine the thickness of steel sheets, paper, aluminium foil and plastic film during production (Figure 4.8.3). The caesium-137 is placed on one side of the material, and a detector, which measures the amount of radiation, is placed on the other. Thicker materials absorb more radiation, so the amount of radiation detected means the material's thickness can be accurately measured.

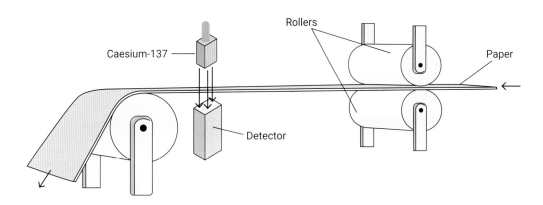

▲ **FIGURE 4.8.3**   Radiation is used to measure the thickness of paper.

Gold-198 is used as a tracer to follow the movement of sewage and other wastes through waterways in a similar way to how radioactive tracers are used in medicine.

Leaks can be detected in underground water pipes or oil lines by adding a radioactive isotope, such as sodium-24, and following the passage of the tracer in the pipe with a detector. The detector will show any leaks as the radioactive isotope escapes the pipes through cracks.

4.8 LEARNING CHECK

1 **Describe** the two uses of medical radioactive isotopes.
2 **Describe** how a radioactive tracer works.
3 **Explain** why medical isotopes generally have short half-lives of hours or days.
4 **Predict** the penetrating power of the sodium-24 isotope used in leak detection in pipes (high, medium or low). **Justify** your prediction.
5 Using one of the examples of radioactive isotopes in industry, **research** the properties of the isotope and **create** a one-slide presentation that links the properties of the isotope to its use.

4.9 First Nations Australians' history

IN THIS MODULE, YOU WILL:
✓ explore dating methods that provide scientific evidence of how long First Nations Australians have lived on the Australian continent.

▲ **FIGURE 4.9.1** Examples of First Nations Australians' ground-edge axes

Joe Sambono

Image courtesy of the River Murray and Mallee Aboriginal Corporation

▲ **FIGURE 4.9.2** Shellfish remains at a midden site. Radiocarbon dating of the shells can indicate how long humans have been in the area.

Radiocarbon dating in Australia

Archaeologists have used radiocarbon dating to study the patterns of movement of humans on the Australian continent. The discovery of human remains (burials) in the remote Willandra Lakes region (New South Wales) has been one of the most important archaeological discoveries. The remains, known as Mungo Man and Mungo Lady, are of immense importance to the Ngiyampaa, the Mutthi Mutthi and the Paakantji Peoples. Radiocarbon dating of the remains estimated them to be at least 25 000 years old. Further analysis of this site using other dating techniques suggests that humans have been in this region for more than 40 000 years. There have been many years of heated debate about the ownership and perceived cultural disrespect of these important human remains. This has included disagreement about whether the remains should or should not be returned to their respective ancestral descendants. The recent reburial of some of these remains has resulted in further debate and pain for some people in the community.

Radiocarbon dating of Mungo Man and Mungo Lady revealed human occupation of the Australian continent for many thousands of years. Findings of artefacts and dating of cave paintings suggest First Nations Australians' occupation of Australia is even longer. Many First Nations Australians believe that their ancestors originated in Australia, on their own clan countries and that First Nations Australians have been on this continent since time immemorial.

The earliest ground-edge axe recovered from Bunuba Country in the south of the West Kimberley region in Western Australia has been dated to about 49 000 years ago. You can see an example of a ground-edge axe in Figure 4.9.1.

Radiocarbon dating of freshwater mussel shells at a midden site on Meru Country in the Pike River region of South Australia showed that they are about 29 000 years old. Shell midden sites are places along waterways where groups of people repeatedly gathered, often using the same site over long periods. The shells come from First Nations Peoples collecting, cooking and eating shellfish. The shells contain carbon in the form of calcium carbonate, which can be extracted for dating (Figure 4.9.2).

Other dating methods

Often, dating of paintings and artefacts relies on a combination of methods. In the northern Kimberly region (Western Australia), mud wasp nests from rock shelters were used to date rock paintings. The nests were built on top of the paintings and so provide some information about the age of the work. These wasp nests were dated using quartz sand grains within the mud. The nests are believed to be up to 18 000 years old, so the paintings underneath them are even older. Figure 4.9.3 shows wasp nests covering rock paintings in Guugu Yimithirr Country, north Queensland.

Radiocarbon dating and sand grain analysis dated tools of the Mirarr People (Kakadu region of the Northern Territory). An excavation uncovered ground-edge hatchets, tools used for seed grinding and ochre crayons used to make pigments (Figure 4.9.4). These artefacts are thought to be at least 65 000 years old.

> **Important!**
>
> Please remember to never touch or disturb **First Nations Australian** artefacts. Report any finds to your local cultural heritage authority.

▲ **FIGURE 4.9.3**   Remains of wasp nests over rock paintings in Guugu Yimithirr Country, north Queensland. By dating the nests, scientists can estimate how old the rock paintings are.

▲ **FIGURE 4.9.4**   Djurrubu Rangers of the Mirarr clan at an excavation site on their Country where tools and pigments were uncovered.

First Nations Peoples' occupation of Australia

> ☆ **ACTIVITY**

1 Use the information in this module to create a table that compares dating studies. Include information such as the:
  a Traditional Owners of the site or name of the Country/Place.
  b material that was used for dating.
  c dating method.
  d age of the material.

2 Why are these studies about First Nations Australians' occupation of Australia important? What have you learned about Australian First Nations Peoples and their respective cultures?

3 How can such studies be done in ways that are respectful to the Traditional Owners of the sites?

4 What are some of the limitations of these methods of dating materials?

4.10 Using radioactive isotopes to track contaminants and pollutants in the environment

BY THE END OF THIS MODULE, YOU WILL BE ABLE TO:

✓ explain how societal values influence scientific research
✓ explain how radioactivity is used to identify and track contaminants in the environment.

Video activity
OPAL nuclear reactor

The Australian Nuclear Science and Technology Organisation (ANSTO) conducts research in a range of science disciplines, including medicine, environmental science and materials development.

ANSTO operates the Open Pool Australian Lightwater (OPAL) nuclear reactor at Lucas Heights in Sydney (Figure 4.10.1). It is a research reactor that produces radioactive isotopes. The OPAL reactor, along with other machines at the facility, can also help to analyse samples of water, soil and air to determine if specific radioactive isotopes are present.

Air, water and soil pollution is a problem in Australia and around the world (Figure 4.10.2). We need accurate information about the location and amount of pollution in the environment (Figure 4.10.3). Farmers need to know that the soil they are growing crops in is not polluted. City gardeners are encouraged to test soils if growing food in case the soil is contaminated by previous land use. People living in cities need to know if air pollution levels are high so they can protect their health. It is vital for everyone that waterways are clean and do not contain toxic contaminants that could get into drinking water or recreational areas.

Image supplied by ANSTO

▲ **FIGURE 4.10.1** The OPAL reactor at Lucas Heights in Sydney

Shutterstock.com/PradeepGaurs

▲ **FIGURE 4.10.2** Air pollution in some cities such as Delhi, India, can be a health hazard to people living there.

Shutterstock.com/APChanel

▲ **FIGURE 4.10.3** A scientist measures levels of radioactive isotopes in the environment.

The scientific team at ANSTO conducts research and monitoring of the environment to:

- monitor changes to the composition of the atmosphere to ensure air quality
- analyse how the soil composition is changing
- analyse water samples to identify any contaminants.

Scientists use radioactive isotopes to:

- identify specific contaminants in soil, air and water so we know what hazards are present (Figure 4.10.4)
- trace the path of nutrients through soils, water through waterways and underground water sources by using radioactive isotopes as trackers
- monitor how pollutant levels change over time by measuring the amount of a known radioactive isotope in the atmosphere, water or soil.

Getty Images/E+/claudio.arnese

▲ **FIGURE 4.10.4** Pollution from industry can affect waterways.

4.10 LEARNING CHECK

1 How does having a nuclear research reactor in Australia help scientists monitor the environment?
2 **Describe** two ways that radioactive isotopes are used to help reduce pollution.
3 **Discuss** (give positives and negatives) the use of radioactive isotopes to monitor the environment.

4.11 Presenting data in tables

SCIENCE SKILLS IN FOCUS

IN THIS MODULE, YOU WILL FOCUS ON LEARNING AND IMPROVING THESE SKILLS:

▶ collecting data from a simulation

▶ using a spreadsheet to present the data in tables and graphs

▶ using provided data to form conclusions, identify trends and make predictions.

PRESENTING DATA IN TABLES

Scientific data is collected in experiments and simulations. To make the best use of this information, it should be presented in a clear, logical and organised way, such as in a table.

You will have constructed tables for previous experiments and will be familiar with using a title, column and row headings, and ruled lines.

Apart from this basic structure, what else can you do to make your table useful and readable?

- **Table layout:** Consider how many columns and rows you need and how wide your columns will be. Generally, you should have more rows than columns, so you don't have thin columns containing text that is difficult to read. Plan your table before you start.

- **Column labelling:** Label columns clearly. 'Trial 1' is not a useful column heading. Try to incorporate what you are measuring; for example, 'concentration of acid' or 'mass of products'.

You should construct your tables before conducting your experiment, so you know exactly what data to measure, with units in header rows. Alternatively, you could use a spreadsheet program such as Excel to construct tables and collect your data.

▼ **TABLE 4.11.1** A poorly constructed table. Rows and columns are not considered, and column headings are not useful if you are not familiar with the experiment

50 g trial 1	50 g trial 2	50 g trial 3	100 g trial 1	100 g trial 2	100 g trial 3	150 g trial 1	150 g trial 2	150 g trial 3

▼ **TABLE 4.11.2** A well-constructed table. Rows and columns have clear headings. Trials and measures of the independent variable are separated

| Mass of calcium carbonate added to hydrochloric acid (g) | Time taken for reaction to complete (s) | | |
	Trial 1	Trial 2	Trial 3
50			
100			
150			

Video
Science skills in a minute: Presenting data in tables

Science skills resource
Science skills in practice: Data tables

INVESTIGATION 1: SIMULATING RADIOACTIVE DECAY

AIM

To model the process of radioactive decay and assess how accurate your model is at showing the process of decay

MATERIALS

- 50 items to model atoms, such as 50 coins, 50 discs, 50 cards or 50 pieces of paper. One side should be different from the other; for example, by putting an X on one side of the paper/disc
- container to shake or shuffle the 50 'atoms'

METHOD

1 Place your 'atoms' into your container and shake the container well for about 10 seconds.

2 Pour out the atoms on a large, flat surface so that they randomly fall with one side facing up.

3 Choose one side (you will stick with this throughout the experiment). For example, if you have discs with an X on one side, pick the X. Remove all atoms with the X. Count how many atoms remain and record this number.

4 Repeat steps 1–3 with the remaining atoms until you have no atoms left.

RESULTS

1 Using the Science skills in focus information, construct an appropriate table to record the number of atoms remaining after each 'shake'. Include the original number of atoms at shake 0.

2 Using Excel, or graph paper, plot a graph of the number of atoms remaining after each 'shake'. Make sure you include the initial number of atoms at 0 shakes. Sketch a curve of best fit to show the trend.

EVALUATION

1 What is represented by the 'atoms' having two sides?

2 Describe how the concept of half-life is represented in this simulation?

3 Examine your graph and data. Using evidence such as the shape of the graph or the number of atoms remaining, describe the trend you observed in this simulation. Include the concept of half-life in your description.

4 Explain two features of this simulation that accurately represent the process of radioactive decay.

5 Explain two features of this simulation that do not accurately represent the process of radioactive decay.

INVESTIGATION 2: DATA ANALYSIS – MAKING PREDICTIONS ABOUT RADIOACTIVITY

AIM

To use tables and graphs to solve problems and make predictions about radioactivity

ACTIVITY 1: FIND THE HALF-LIFE

Use Table 4.11.3 to determine the half-life of the element. Hint: you should use graph paper or Excel to draw a decay curve as your first step.

▼ TABLE 4.11.3   Activity of a radioactive isotope

Time (s)	Activity (counts)
0	7112
84	6650
220	5964
346	5392
467	4894
600	4400

1   Justify the method you used to determine the half-life.

2   Describe two limitations you encountered while determining the half-life that reduce the reliability of the answer.

ACTIVITY 2: PREDICTING THE BEST RADIOACTIVE ISOTOPE TO USE

The radioactive isotopes in Table 4.11.4 include three that are used for medical purposes and two that are used in industry.

▼ TABLE 4.11.4   Some isotopes, their half-lives and the radiation they produce

Isotope	Half-life	Type of radiation produced
Iodine-123	13 hours	Gamma
Cobalt-57	272 days	Gamma
Thallium-201	73 hours	Gamma
Americium-241	432 years	Alpha and gamma
Sodium-24	15 hours	Beta

1   a   Identify two isotopes from the table that could be used for medical purposes.

   b   Justify your answer using two properties of isotopes used for medicine.

2   Sodium-24 is used in industry for tracking the movement of water through water pipes around houses. It has a short half-life. Suggest why an isotope is used that only has a very short half-life.

3   Justify which of the five isotopes is not suitable for use in medicine.

1 Copy and complete the following table to summarise the different atomic models.

Scientist	Diagram of atomic model	Summary of model
Dalton		
		Negatively charged particles in a sphere of positive charge
	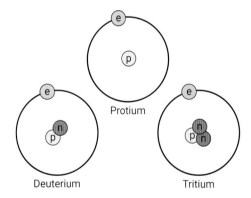 Nucleus, Orbit, Electron	
Bohr		

2 **Identify** the number of protons and neutrons in the following elements.

a $^{27}_{13}\text{Al}$ b $^{65}_{30}\text{Zn}$

3 **Compare** the number of subatomic particles in the isotopes oxygen-16 and oxygen-18. The atomic number of oxygen is 8.

4 **Explain** why radioactive decay occurs in radioisotopes.

5 **Compare** the structure of alpha particles, beta particles and gamma radiation.

6 **Describe** three different uses of radioisotopes. Include one medical and one industrial example in your response.

UNDERSTANDING

7 **Explain** how Thomson confirmed the existence of subatomic particles in the atom.

8 The notation ^A_ZX is used to represent atomic structure. **Explain** why the notation of hydrogen is unique among all the elements.

9 Carbon-14 has a half-life of 5730 years. **Explain** what information you can determine from this statement.

10 **Describe** the forces that act in the nucleus of atoms and **explain** why some isotopes are unstable.

11 **Explain** the key property shared by all medical radioisotopes used in the human body.

APPLYING

12 When creating artificial isotopes, less energy is required for neutron bombardment than for proton bombardment. **Explain** this statement in terms of the charge on the particles **involved**.

13 The following diagram shows three isotopes of a common element. Use the diagram to write the correct notation for each of the three isotopes in the format 'element-X' (e.g. oxygen-16).

Protium

Deuterium

Tritium

14 Scientists who work with alpha radiation wear only gloves to protect themselves. Scientists working with beta radiation often work with robotic arms with the radiation source contained within a metal-lined box. **Explain** why the precautions differ with different types of radiation sources.

ANALYSING

15 **Suggest** reasons why Rutherford's gold foil experiment showed the presence of a nucleus in the middle of the atom rather than the positive charge being spread across the whole atom.

16 The following graph shows the decay curve for a radioisotope. The measurement of radiation on the vertical axis is 'counts per minute', which is how much radiation is recorded on a detector.

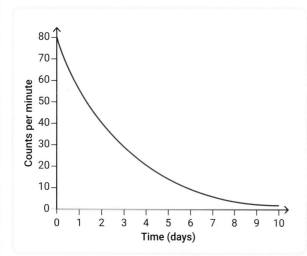

a Using the graph, **determine** the half-life of this radioisotope.

b **Determine** the counts per minute that would be recorded after three half-lives.

17 The radioisotope polonium-208 has 84 protons. By comparing the products of the beta decay and alpha decay of polonium-208, **explain** why this radioisotope undergoes alpha decay.

18 Carbon-14 dating is not considered accurate for objects less than 4000 years old or more than 60 000 years old. Explain why, by considering the half-life of carbon-14.

19 **Explain** how dating the remains of wasp nests can be used to determine the age of First Nations Australian rock paintings.

20 **Sketch** a radioisotope decay curve and use it to **explain** why it is difficult to predict masses after more than five half-lives have passed.

21 Scientists have found fossilised skulls of hominids (human ancestors) embedded in rocks estimated to be 2–5 million years old. This means that the hominids died at the same time as the rocks formed, so they are the same age. Evaluate the pros and cons of using carbon-14 dating and uranium–lead dating to determine the age of the skulls.

22 A scientist determines that a rock has equal amounts of uranium and lead and thus states one half-life has passed and the rock is 710 million years old. **Assess** the accuracy of this statement. If there is information missing, rewrite the statement correctly and **justify** your answer.

23 **Create** a flow chart to show the development of the atomic model. You should include the scientists involved, how they confirmed their theories, and how it advanced the previous model.

24 **Create** a decay curve graph for carbon-14 that you could use to estimate the age of materials. Include an explanation of how you could use the decay curve. Ask a classmate to try it out with an example (make up one based on your decay curve). Get feedback on your explanation. Make improvements based on the feedback until you have an explanation a classmate can use.

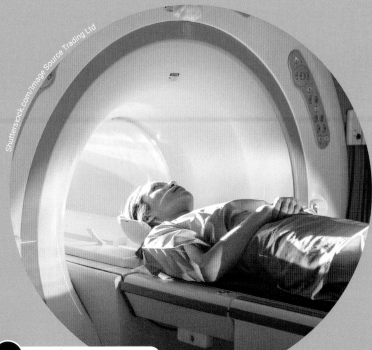

1 Connect what you've learned

In this chapter, you have learned about the structure of the atom, and how it can change through radioactive decay.

Draw a mind map to show how the structure of the atom links to the different types of radiation. Use the key words from each module to prompt you.

2 Check your thinking

At the start of the chapter, you were asked 'What have TV, books, movies or comic books told you about radiation?' and to reflect on how the image of the radioactive warning signs made you feel. How have your feelings about the warning signs changed? If they haven't, why do you think you still feel that way?

At the start of the chapter, you were asked if you knew any examples of how radiation is used. How has your understanding of the uses of radiation changed?

3 Make an action plan

What do you think the view of wider society on radiation is? How do you think you may be able to educate people on the science involved? Have a discussion with your parents about their knowledge of radiation, and the advantages and disadvantages of radioactivity.

4 Communicate

Make a poster or presentation showing why radioactivity is useful and how it affects your everyday life.

5 Chemical reactions and equations

5.1 **Chemical reactions** (p. 140)

Chemical reactions involve rearranging atoms of reactants to form new substances, with energy being released or absorbed.

5.2 **Word equations** (p. 144)

Word equations show the reactants and products in a chemical reaction.

5.3 **Law of conservation of mass** (p. 147)

In chemical reactions, mass is conserved so the masses of reactants and products are equal.

5.4 **Balanced chemical equations** (p. 150)

Balanced chemical equations show chemical formulas and obey the law of conservation of mass.

5.5 **Chemical reactions in our lives: acids and bases** (p. 154)

The acid or base levels in a solution can be measured by using indicators, or with the pH scale.

5.6 **Chemical reactions in our lives: corrosion** (p. 158)

Corrosion is a chemical process that often happens to metals exposed to water and oxygen.

5.7 **Chemical reactions in our lives: reactions of bioethanol** (p. 161)

The production and use of bioethanol involves a series of chemical reactions including photosynthesis, fermentation and combustion.

5.8 FIRST NATIONS SCIENCE CONTEXTS: **First Nations Australians' chemistry to develop pigments and dyes** (p. 164)

Investigate the ways First Nations Peoples developed pigments and dyes through understanding of chemical reactions.

5.9 SCIENCE AS A HUMAN ENDEAVOUR: **Bioethanol as a more environmentally friendly fuel** (p. 168)

Bioethanol is used in vehicle fuel in many countries.

5.10 SCIENCE INVESTIGATIONS: **Controlling variables to perform a valid investigation** (p. 171)

1 How does the amount of sugar affect the rate of production of ethanol?
2 Do all fuels produce the same amount of energy?

▲ **FIGURE 5.0.1**   Bioethanol contains ethanol, which comes from plants.

You may have travelled in a vehicle that uses food as fuel. In Australia bioethanol is available in many petrol stations as E10 fuel. This is regular petrol with about 10 per cent ethanol added to it. Worldwide, more than 60 countries use bioethanol in fuel in an effort to use cleaner energy. In Australia, not all states have widely adopted the use of E10 fuel yet.

▶ How does food become part of petrol?

▶ Is E10 fuel sold in petrol stations in your state?

▶ Why would using crops and food make the energy from petrol cleaner?

▶ Can you see any problems with using food and crops as a fuel?

#5 SCIENCE CHALLENGE ACCEPTED!

At the end of this chapter, you can complete the Big Science Challenge Project #5. You can use the information you learn in this chapter to complete the project.

Assessments

- Prior knowledge quiz
- Chapter review questions
- End-of-chapter test
- Portfolio assessment task: Science investigation

Videos

- Science skills in a minute: Controlling variables **(5.10)**
- Video activities: Chemical reactions in 'food comas' **(5.1)**; What is the law of conservation of mass? **(5.3)**; What are acids and bases? **(5.5)**; Oxidation reactions **(5.6)**; Biofuels **(5.7)**; Oxygen and combustion **(5.7)**; Algae as a biofuel **(5.9)**

Science skills resources

- Science skills in practice: Controlling variables for a valid investigation **(5.10)**
- Extra science investigations: Comparing mass loss in chemical reactions **(5.3)**; Neutralisation reactions **(5.5)**; Making sherbet **(5.5)**

Interactive resources

- Simulations: Balancing chemical equations **(5.4)**; pH scale **(5.5)**
- Match: Word equations **(5.2)**
- Drag and drop: Signs of a chemical reaction **(5.1)**
- Crossword: Chemical reactions **(5.7)**

❈ Nelson MindTap

To access these resources and many more, visit:
cengage.com.au/nelsonmindtap

Video activity
Chemical reactions in 'food comas'

Interactive resource
Drag and drop:
Signs of a chemical reaction

physical change
a change of state or appearance of a substance or an object without a change in its chemical composition

chemical reaction
a process involving the rearrangement of atoms in reactants to form new chemical products, through a chemical change

chemical change
a change that results in the formation of a new substance in a chemical reaction

reactant
an initial chemical in a chemical reaction

product
a chemical formed in a chemical reaction

molecule
two or more non-metal atoms bonded together

GET THINKING

Look at Figures 5.1.1 and 5.1.2. They both represent a chemical reaction. What do you think is occurring in these reactions? Which parts of the diagrams give you information? Consider how you might present information in diagram form as part of your written work or notes.

What is a chemical reaction?

You may remember that a **physical change** is when a substance or an object changes without altering its chemical composition. An example of a physical change is a change of state such as freezing or evaporating, or a process such as dissolving. A **chemical reaction** occurs when chemical substances react, and a **chemical change** occurs. This means one or more new substances has formed with a different chemical composition from the original substances. A chemical change can be identified by a range of observations, including:

- change of colour
- formation of gas (bubbles)
- formation of solid
- production of light or heat energy.

A chemical reaction starts with substances called **reactants**. The final substances produced in the chemical reaction are called **products**. In Figure 5.1.1, the oxygen and hydrogen gas are the reactants, and they undergo chemical change during the reaction to form the product, water.

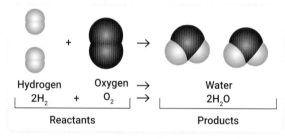

Hydrogen		Oxygen	→	Water
$2H_2$	+	O_2	→	$2H_2O$
Reactants				Products

▲ **FIGURE 5.1.1** The reaction of oxygen and hydrogen to form water is a chemical reaction.

What happens to atoms in a chemical reaction?

During a chemical reaction, the atoms rearrange to form new substances. In chemical reactions, the number and type of atoms stay the same from the reactants to the products. The atoms are only rearranged. In Figure 5.1.1, you can see that the reactants are two hydrogen **molecules** (two hydrogen atoms) and an oxygen molecule (two oxygen atoms). Molecules are structures formed by two or more non-metal atoms.

Count the number of hydrogen and oxygen atoms in the reactants. Now count the atoms in the products. What do you notice? The two sides have the same number of the same type of atom. In this case, both reactants and products have 4 hydrogen atoms and 2 oxygen atoms. That's because in a chemical reaction, it is only the arrangement of atoms that changes.

Energy in chemical reactions

When atoms rearrange and new chemical substances form, bonds must be broken in the reactants and new bonds formed in the products (Figure 5.1.2). Chemical energy is a form of potential energy. This energy is stored in the bonds that join atoms. Chemical reactions can allow this potential energy to be released to the surroundings as heat or light. You will learn more about energy in Chapters 7 and 8.

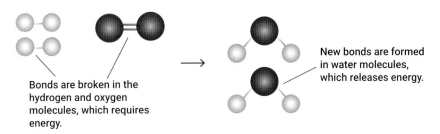

Bonds are broken in the hydrogen and oxygen molecules, which requires energy.

New bonds are formed in water molecules, which releases energy.

▲ FIGURE 5.1.2   Bond making and breaking in the formation of water

In every chemical reaction:

* bonds are broken in the reactants to separate atoms – this needs energy from the surroundings
* bonds are formed in the products to join atoms – this releases energy to the surroundings.

This means that every chemical reaction uses energy to break down reactants and releases energy when products are formed. In the graph of the reaction in Figure 5.1.3a, you can see that more energy is released in bond making than is used in bond breaking. Overall, this means that this reaction releases energy. Chemical reactions that release energy are called **exothermic** reactions.

The graph in Figure 5.1.3b shows an **endothermic** reaction in which the energy used is greater than the energy released. Overall, energy is absorbed by this reaction. Chemical reactions that absorb energy are called endothermic reactions.

exothermic
a chemical reaction that releases energy

endothermic
a chemical reaction that absorbs energy

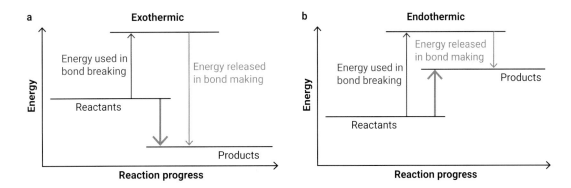

▲ FIGURE 5.1.3   (a) An exothermic reaction releases energy. (b) An endothermic reaction absorbs energy. In both graphs, the thin red arrow shows energy used, the thin blue arrow shows energy released and the thick blue arrow shows the overall energy released or used in the reaction.

Examples of chemical reactions

Almost everything that you come into contact with was made by a chemical reaction. You observe hundreds of reactions a day; they even occur constantly inside your body!

One of the places you are most likely to encounter a chemical reaction is food. Most food is made through chemical reactions. The baking powder in bread, cakes, biscuits and pastries reacts to produce carbon dioxide, which makes them rise (Figure 5.1.4). Cooking food such as eggs or steak causes sugars and proteins to break down and form new compounds with different flavours. Even a salad begins with a chemical reaction. Plants form sugars through a chemical reaction called photosynthesis. We will look at this process later in this chapter and when learning about the carbon cycle in Chapter 6.

Shutterstock.com/Praisaeng

▲ **FIGURE 5.1.4** Baking involves chemical reactions.

All medicines are produced through carefully controlled chemical reactions. An example is the production of aspirin, used to treat headaches, fever and pain. Aspirin was developed in laboratories by copying a chemical found in willow tree bark. Most medicines are very complex molecules, as seen in Figure 5.1.5.

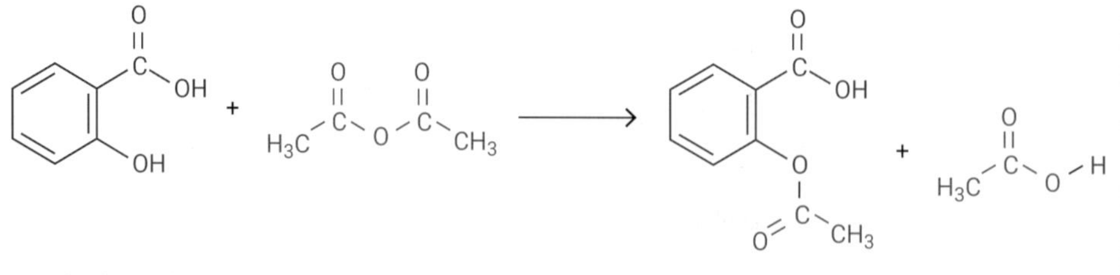

Salicylic acid
$C_7H_6O_3$

Acetic anhydride
$C_4H_6O_3$

Acetylsalicylic acid
$C_9H_8O_4$

Acetic acid
$C_2H_4O_2$

▲ **FIGURE 5.1.5** The chemical reaction that produces aspirin is very complex. The scientific name for aspirin is acetylsalicylic acid.

Other chemical reactions you might be familiar with are the:

- discharging and recharging of batteries in your phone and laptop
- production of plastics, polymers and most fibres you wear (e.g. nylon)
- production of concrete, paints, glues and many other building and construction materials.

Observing a chemical reaction – elephant's toothpaste

Follow the instructions below to set up a spectacular chemical reaction. In this reaction the reactant is hydrogen peroxide (containing hydrogen and oxygen). The products are water and oxygen. The experiment also uses yeast as a catalyst. A catalyst is a chemical that helps speed up a chemical reaction.

This experiment can be messy. It is recommended that you do it outside, in a large sink or trough, or on a tarpaulin.

What to do

- Pour 125 mL of 6 per cent hydrogen peroxide into a clean, dry 1.25 L plastic drink bottle.
- Add 1–2 mL of dishwashing detergent (to create bubbles) and 1 mL of food colouring.
- Place a sachet of yeast (e.g. baker's yeast) into a 250 mL beaker and add 50 mL of warm water. Stir for 30 seconds.
- Place the 1.25 L bottle onto a tarpaulin, or in a sink to catch the overflow. Place a funnel in the top of the bottle.
- Pour the yeast solution into the bottle, quickly remove the funnel and move away!

What do you think?

1. A molecule of hydrogen peroxide has two hydrogen and two oxygen atoms. A molecule of oxygen has two atoms of oxygen. Water has an oxygen bonded to two hydrogen atoms. **Describe** what bonds need to be broken and formed in this chemical reaction.

2. What observations did you make that tell you this was a chemical reaction?

3. At the end of the reaction, the bottle would have been warm. **Identify** the reaction as endothermic or exothermic.

4. Based on your answer to Question 3, **explain** whether more energy was absorbed to break bonds, or more energy was released when forming bonds in this reaction.

5.1 LEARNING CHECK

1. **Identify** four observations that provide evidence of a chemical reaction occurring.

2. A chemical reaction needs more energy to break down the reactant molecules than to form the product molecules. **Justify** whether the reaction is endothermic or exothermic.

3. **Describe** five chemical reactions you have already come into contact with, seen or used today. Consider what you have eaten, how you got to school, what you are wearing or what you are using.

Interactive resource
Match: Word equations

Features of chemical equations

chemical equation
a way of presenting information about chemical reactions

A **chemical equation** is a way of presenting information about chemical reactions.

In Module 5.1, we looked at the chemical reaction involving hydrogen and oxygen forming water. The word equation for this reaction is:

$$\text{hydrogen} + \text{oxygen} \rightarrow \text{water}$$

A chemical equation has the following features.

- All reactants are on the left side of the equation.
- All products are on the right side of the equation.
- An arrow points from the reactants to the products.
- If there is more than one reactant or product, a plus sign (+) is used between the chemicals; for example, 'hydrogen + oxygen'.

Showing conditions in chemical equations

A chemical equation shows the chemicals involved in the reaction. Sometimes other factors are involved that are not chemicals. For example, some chemical reactions need heat or energy to start them, or they produce energy (Figure 5.2.1). Other factors that affect chemical reactions are pressure or use of a catalyst to speed up the reaction, as you saw in the elephant's toothpaste activity in Module 5.1. So how can we show these other factors? We call these other factors **conditions** and can put them over the arrow in the equation:

conditions
factors that affect a chemical reaction, such as heat or pressure

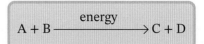

$$A + B \xrightarrow{\text{energy}} C + D$$

► **FIGURE 5.2.1**
A sparkler needs energy to start a chemical reaction.

Describing the state of a chemical

Sometimes it is helpful to identify the state of a chemical. Water can exist as a solid (ice), a liquid (water) and a gas (steam) (Figure 5.2.2). When water is a reactant or product, sometimes it is a liquid, and sometimes a gas. Instead of writing 'water gas' or 'steam' (steam is not a chemical name), you can use state symbols.

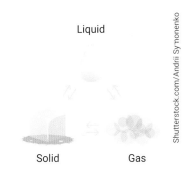

Liquid

Solid   Gas

▲ **FIGURE 5.2.2**   Water can exist
as a solid, a liquid and a gas.

State symbols are written in brackets after a substance to show whether it is solid (s), liquid (l), gas (g) or dissolved in water as a solution (aq). The 'aq' for a solution represents **aqueous**. An aqueous solution is a substance that is dissolved in water. Pure water with nothing dissolved in it is not aqueous; it is a liquid.

aqueous
a solution containing a substance dissolved in water

$$\text{hydrogen (g) + oxygen (g)} \rightarrow \text{water (l)}$$

Steps in writing a word equation

The steps to writing a word equation are as follows.

1 Write the reactants with a + sign between each reactant.

2 Put an arrow after the reactants, then write the products with a + sign between each product.

3 Put any conditions required over the arrow.

4 Include state symbols for each reactant and product.

Writing word equations from chemical experiments

☆ **ACTIVITY**

You will perform four chemical reactions and practise writing word equations. You will also make observations that show a chemical reaction has occurred.

For each reaction:

• conduct the experiment as described

• use the information provided to write a word equation, including states and any other conditions required

• record observations that show a chemical reaction has occurred

 Warning

• Chemicals can cause skin irritation or more severe problems if swallowed or they get into your eyes. Wear appropriate PPE (gloves, safety glasses and lab coat) to protect yourself.

• Check with your teacher about how to dispose of the chemicals you use.

Observations are what you can see, not conclusions or assumptions. So, you cannot observe 'hydrogen gas forming', you can only observe 'bubbles of gas forming'.

Follow the instructions below, record your results, and then answer the questions.

What to do

Chemical reaction 1: magnesium metal and hydrochloric acid

- In a 100 mL conical flask, add 20 mL of 1.0 mol L^{-1} hydrochloric acid solution to a 3 cm strip of magnesium metal.
- The products of this reaction are magnesium chloride solution and hydrogen gas.

To give you another example of a chemical equation, reaction 1 is shown below:

Magnesium (s) + hydrochloric acid (aq) → magnesium chloride (aq) + hydrogen (g)

Chemical reaction 2: copper sulfate solution and sodium hydroxide solution

- In a micro test tube (or the smallest test tube you have) add 10 drops of 0.25 mol L^{-1} copper sulfate solution to 10 drops of 0.25 mol L^{-1} sodium hydroxide solution.
- The products of this reaction are copper hydroxide solid and sodium sulfate solution.

Chemical reaction 3: hydrochloric acid and sodium hydroxide

- In a micro test tube (or the smallest test tube you have) add 10 drops of 0.25 mol L^{-1} hydrochloric acid solution to 10 drops of 0.25 mol L^{-1} sodium hydroxide solution.
- The products of this reaction are sodium chloride solution and liquid water.

Chemical reaction 4: methane and oxygen

- Light a Bunsen burner and turn it to a blue flame. The Bunsen burner gas is methane, and it burns in the oxygen in the air.
- The products of this reaction are carbon dioxide and water.

What do you think?

1 Did all the chemical reactions show obvious signs of a chemical change occurring? Use specific examples from the four experiments in your answer.

2 Did any of the four reactions you performed require heat to start?

5.2 LEARNING CHECK

1 **Identify** the different states that can be included in a chemical equation and why they might be necessary.

2 How are conditions such as temperature and pressure represented in a chemical equation? Why is this?

3 **Write** word equations for the following. Include states and reaction conditions if they are appropriate.

  a Solid pieces of calcium react with oxygen gas when heated to form solid calcium oxide.

  b Hydrogen peroxide (liquid bleach) decomposes when heated to form oxygen gas and liquid water.

  c Copper nitrate solution and sodium hydroxide solution combine to form solid copper hydroxide and a solution of sodium nitrate.

9780170472838

5.3 Law of conservation of mass

BY THE END OF THIS MODULE, YOU WILL BE ABLE TO:

✓ explain how the law of conservation of mass relates to chemical reactions.

This module is about a law in chemistry. What is your understanding of a law in science? How can laws be used?

Video activity
What is the law of conservation of mass?

Extra science investigation
Comparing mass loss in chemical reactions

law of conservation of mass
the total mass of reactants and products in a chemical reaction is equal

What is the law of conservation of mass?

In the 1700s, the French scientist Antoine Lavoisier conducted experiments and proposed a theory about chemical reactions. After many careful observations, he noticed that the mass of the reactants in a reaction was always the same as the mass of the products.

This became known as the **law of conservation of mass**:

> Total mass of reactants = total mass of products

Figure 5.3.1 shows an example you might be familiar with. When you watch a campfire burning in winter you are observing a chemical reaction. This is represented as:

Wood + oxygen → carbon dioxide + carbon + water

Wood contains several substances that burn in oxygen from the air. The products of this reaction are ash (carbon), carbon dioxide gas and water vapour. The combined reactants – wood and oxygen – have a certain mass. If you were to measure the mass of the carbon, carbon dioxide and water formed, it would be the same as the mass of the combined reactants, even though you might not be able to see or capture them easily.

Reactants – wood and oxygen

Reaction occurring – combustion of wood

Products – ash and gases released

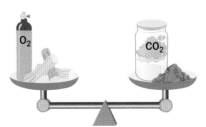

Mass of reactants = mass of products

▲ **FIGURE 5.3.1** Conservation of mass when burning firewood

How does conservation of mass apply to observations in a chemical reaction?

Figure 5.3.2 shows an experiment in which silver nitrate solution is added to sodium chloride solution. If you were to conduct the experiment as shown, you would record the same mass before and after the reaction. In the first image, the chemicals are separated; the silver nitrate is in the test tube separate from the sodium chloride in the conical flask.

When the test tube is inverted, the silver nitrate pours out and mixes with the sodium chloride. The products formed are a white silver chloride solid, and a sodium nitrate solution.

Silver nitrate (aq) + sodium chloride (aq) → silver chloride (s) + sodium nitrate (aq)

Note that the mass of the reactants is the same as the mass of the products.

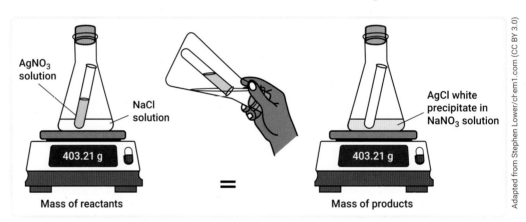

▲ FIGURE 5.3.2   The chemical reaction between silver nitrate ($AgNO_3$) solution and sodium chloride (NaCl) solution

---

☆ ACTIVITY

Conservation of mass

You will perform a simple conservation of mass experiment. Follow the instructions below, record your results, then answer the questions.

What to do

- Measure 20 mL of potassium iodide solution and pour it into a 50 mL beaker.
- Measure 20 mL of lead nitrate solution and pour it into a different 50 mL beaker.
- Place both beakers on an electronic balance at the same time and record the initial mass of the beakers and reactants.
- Pour one beaker into the other and replace the empty beaker back on the electronic balance.
- Record the final mass and any observations that show a chemical change.

What do you think?

1 Did this experiment demonstrate the law of conservation of mass? **Justify** your answer.

2 **Explain** why the empty beaker needed to be placed on the electronic balance. What would have happened to the results if it had been left on the bench?

3 **State** at least two observations that could be used as evidence for a chemical reaction.

9780170472838

1 When solid magnesium is added to copper sulfate solution, as shown in Figure 5.3.3, solid copper and magnesium sulfate solution are formed.

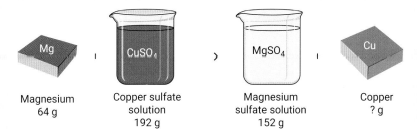

Magnesium
64 g

Copper sulfate
solution
192 g

Magnesium
sulfate solution
152 g

Copper
? g

▲ **FIGURE 5.3.3**　The chemical reaction between magnesium and copper sulfate solution

   a **Write** a word equation to represent this reaction.

   b Use the law of conservation of mass to **predict** the mass of copper that will form.

2 Burning a candle is a chemical reaction (Figure 5.3.4). Wax burns in oxygen to form carbon dioxide and water vapour.

   a **Write** a word equation to represent this reaction.

   b The mass of the candle at the start does not equal the mass of the candle at the end. **Explain** why this does not make the law of conservation of mass wrong. Consider the nature of the products formed in the reaction in your answer.

▲ **FIGURE 5.3.4**　Burning a candle is a chemical reaction.

BY THE END OF THIS MODULE, YOU WILL BE ABLE TO:

✓ write a balanced chemical equation to represent a chemical reaction when given the formula of reactants or products

✓ explain why a chemical reaction needs to be balanced by using the law of conservation of mass.

Interactive resource
Simulation: Balancing chemical equations

GET THINKING

This module uses a lot of chemical formulas. A common formula you might know is H_2O. Brainstorm and write down as many chemical formulas that you have heard of. Think about what all the letters and numbers mean, then write down what you think a chemical formula represents.

Chemical formulas

So far, we have looked at word equations to represent chemical reactions. Chemists have a system of representing chemicals that describes their structure and composition. Some of these you will already be familiar with.

chemical formula
symbols and numbers used to represent the composition of a chemical

H_2O is the **chemical formula** for water. This tells you it has two hydrogen atoms and one oxygen atom in its structure. CO_2 is another you might be familiar with. A carbon dioxide molecule has one carbon atom and two oxygen atoms.

In this chapter, you will be given any formulas that you need.

Balanced chemical equations

The word equations you learned about in Module 5.2 have some uses, but don't tell us a lot of information about the chemicals involved in the chemical reaction.

balanced chemical equation
a chemical equation in which there are the same number of atoms of each element on each side

Table 5.4.1 shows some of the word equations we have already looked at, and what they look like as **balanced chemical equations**.

▼ **TABLE 5.4.1**   Word equations and corresponding balanced chemical equations

Word equation	oxygen + hydrogen → water
Chemical equation	$O_2(g) + 2H_2(g) \rightarrow 2H_2O(l)$
Word equation	magnesium (s) + hydrochloric acid (aq) → magnesium chloride (aq) + hydrogen (g)
Chemical equation	$Mg(s) + 2HCl(aq) \rightarrow MgCl_2(aq) + H_2(g)$
Word equation	wax + oxygen → carbon dioxide and water
Chemical equation	$C_{23}H_{48}(s) + 35O_2(g) \rightarrow 23CO_2(g) + 24H_2O(l)$

What do balanced chemical equations show?

- The number of atoms in a structure – an O_2 molecule has two oxygen atoms, $MgCl_2$ has one magnesium atom for every two chlorine atoms.

- The number of each type of structure that is required – one O_2 molecule reacts with two H_2 molecules to form two H_2O molecules.

Law of conservation of mass

The numbers in front of the chemical formulas are called **coefficients**. They show the ratio of the chemicals that are needed to react. You should recall from Module 5.3 that mass is conserved in chemical reactions. If we look at this statement in terms of the atoms in a chemical reaction, it tells us that you must have the same number and type of each atom in the reactants and in the products.

coefficient
a number placed in front of a chemical formula to balance an equation

Look at the reaction $2H_2(g) + O_2(g) \rightarrow 2H_2O(l)$ shown in Figure 5.4.1.

Table 5.4.2 shows the number of atoms present in the chemical reaction between hydrogen and water. The number of oxygen atoms is the same in the reactants and products. The number of hydrogen atoms is also the same in the reactants and products.

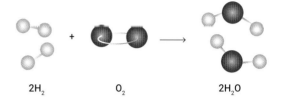

$2H_2$ O_2 $2H_2O$

▲ **FIGURE 5.4.1** The chemical reaction between hydrogen and oxygen to form water

▼ **TABLE 5.4.2** Atoms present in the reaction between hydrogen and oxygen to form water

	Reactants	Products
Formulas	$O_2 + 2H_2$	$2H_2O$
Number of oxygen atoms	2 – there is one O_2 molecule, which has two oxygen atoms – so two in total	2 – each H_2O molecule has one oxygen atom – so two in total
Number of hydrogen atoms	4 – there are two H_2 molecules, each of which has two atoms, so $2 \times 2 = 4$ or four in total	4 – each H_2O molecule has two hydrogen atoms – so four in total

The coefficients are used to make sure the number of each type of atom is equal on both sides of the arrow.

How to balance equations

Figure 5.4.2 shows balanced and unbalanced equations. It uses the example of the addition of hydrogen and oxygen to produce water.

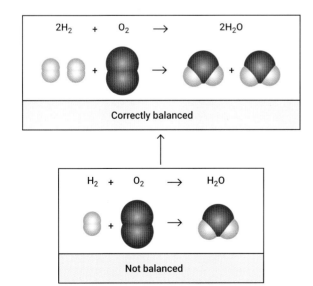

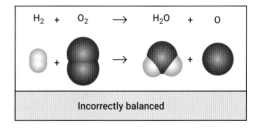

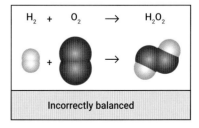

▲ **FIGURE 5.4.2** Equations must be balanced in the correct way; for example, you should not change a chemical formula in order to balance the equation.

Table 5.4.3 shows the steps involved in balancing an equation. In Step 3, the instruction says to not change formulas. This is very important. For example, we want to balance the equation $O_2 + H_2 \rightarrow H_2O$. It is tempting to just add an oxygen to the product to balance the equation. This would give us a product of H_2O_2 and the equation would be balanced.

But when you change the formula, you can often end up with a very different chemical. H_2O is water. H_2O_2 is hydrogen peroxide, or bleach – a very different substance!

Another example of this is O_2 and O_3. With only one oxygen atom different, how different can they be? O_2 is the oxygen gas we need to breathe to survive. O_3 is ozone, a toxic pale blue gas.

So, we only balance equations by placing coefficients in front of the chemicals in the equation.

▼ TABLE 5.4.3   Steps to balancing an equation

Step	Example
1 Write the correct chemical formula for each reactant and product.	$CH_4 + O_2 \rightarrow CO_2 + H_2O$
2 Count the number of atoms of each type on both sides of the arrow (both sides of the reaction).	• 1 atom of C on each side • 4 atoms of H on the left, 2 atoms of H on the right • 2 atoms of O on the left, 3 atoms of O on the right
3 If the numbers on each side are not the same, place numbers in front of the formulas until they are balanced. You may need to go back and forth a few times until you get the numbers right. *Do not change any of the formulas!* • The numbers in front apply to all atoms in the formula. *Do not take numbers away!* • You can only balance equations by adding numbers in front of the formula. You can't take numbers away.	• Place a 2 in front of H_2O to get $2H_2O$. • This makes 4 atoms of H on each side (2 molecules with 2 H in each molecule – $2 \times 2 = 4$ H). • Place a 2 in front of the O_2 to get $2O_2$. • This makes 4 O on each side (2 molecules with 2 atoms of O in each – $2 \times 2 = 4$ O). Balanced equation: $CH_4 + 2O_2 \rightarrow CO_2 + 2H_2O$
4 Do a final check to make sure the equation is balanced.	• 1 atom of C on each side • 4 atoms of H on each side • 4 atoms of O on each side

☆ ACTIVITY

Modelling chemical equations

You will need some plasticine of different colours or a chemical modelling kit. You are going to model chemical formulas and use the model to help you balance the equation. Follow the instructions listed below to show you how this works.

1 $N_2 + H_2 \rightarrow NH_3$ (nitrogen + hydrogen → ammonia)

- Make a nitrogen molecule (N_2) by making two balls of colour 1 and placing them next to each other or joining them with a toothpick.
- Make a hydrogen molecule (H_2) by making two balls of colour 2 and placing them next to each other or joining them with a toothpick.
- Make an ammonia molecule (NH_3) by making one ball of colour 1 and three balls of colour 2 and joining them so they look exactly like the structure in Figure 5.4.3.

▷

- Decide how many of each molecule you need to make so that the nitrogen and hydrogen atoms are balanced in the equation.
- Make all the molecules you need until the atoms are balanced.
- Write the balanced equation.

2 Repeat the modelling process to balance the following equations. The structure of unknown molecules is not important here, just the number of atoms in the molecule.

a $C_2H_4 + O_2 \rightarrow CO_2 + H_2O$

b $N_2 + O_2 \rightarrow NO_2$

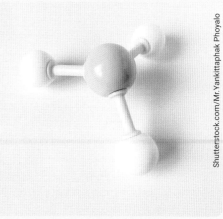

▲ FIGURE 5.4.3   A model of an ammonia molecule

5.4 LEARNING CHECK

1 For each of the following equations, complete a table similar to one shown here. You can use Table 5.4.2 to help guide you.

	Reactants	Products
Formula		
Number of atom type 1		
Number of atom type 2		

a $Mg(s) + 2HCl(aq) \rightarrow MgCl_2(aq) + H_2(g)$

b $C_{23}H_{48}(s) + 35O_2(g) \rightarrow 23CO_2(g) + 24H_2O(l)$

2 Balance the following equations. Use the modelling approach from the activity in this module if necessary or try to balance the equations without making a model.

a $C_3H_8 + O_2 \rightarrow CO_2 + H_2O$

b $Fe + HCl \rightarrow H_2 + FeCl_2$

c $P + O_2 \rightarrow P_4O_{10}$

d $HCl + CaCO_3 \rightarrow CaCl_2 + CO_2 + H_2O$

BY THE END OF THIS MODULE, YOU WILL BE ABLE TO:
- ✓ describe properties and everyday uses of acids and bases
- ✓ explain what pH shows about acidic and basic solutions
- ✓ use information about indicators to identify acidic and basic solutions.

Video activity
What are acids and bases?

Interactive resource
Simulation: pH scale

GET THINKING

Examine the pictures of indicators in this module. Colours are important in chemistry to identify substances and perform tests. Think about how colours can give you information in everyday life. One example is traffic lights. What others can you think of?

What is an acid?

Acids are commonly occurring substances. The vinegar you use as a salad dressing or for cooking contains an acid called ethanoic acid (Figure 5.5.1). Your body produces hydrochloric acid to help digest food. Citrus fruits such as oranges, limes and lemons contain citric acid. The burning sensation you feel in your muscles after intense exercise is due to lactic acid. In Science classes, you have probably performed experiments with acids such as hydrochloric acid.

acid
a substance that can donate a hydrogen ion when in solution

Acids have common physical properties. All acids:
- taste bitter (Note: You will know the taste of lemon and orange juice, but *never* taste acids or any food in a laboratory.)
- are soluble in water
- conduct electricity when in solution.

Acids and bases involve the formation of **ions**. Ions are charged particles that form when atoms gain or lose electrons. For example, a **hydrogen ion** forms when a hydrogen atom (1 proton and 1 electron) loses its electron. When it does this, it has an overall charge because the number of protons and electrons is no longer balanced.

▲ FIGURE 5.5.1   Vinegar contains ethanoic acid.

Shutterstock.com/focal point

ion
a charged particle that forms when an atom gains or loses electrons

hydrogen ion
a hydrogen atom that has lost its electron; represented by H⁺

Acids have different chemical structures, but always have one thing in common – acids produce hydrogen ions when dissolved in water. This means that all acids contain hydrogen. For example, when hydrochloric acid (HCl) is in solution, there are hydrogen ions (H^+) and chloride ions (Cl^-).

What is a base?

You can find **bases** in your home. Bases such as sodium bicarbonate are used for cooking, whereas bleach and sodium hydroxide occur in many cleaning materials (Figure 5.5.2). You might be familiar with bases in the school laboratory, including carbonates, hydroxides and oxides.

Bases are bitter and feel slippery or soapy to touch. Like acids, bases conduct electricity when in solution.

Neutralisation

The chemical reaction between an acid and a base is called neutralisation. When a base dissolves in water, it produces hydroxide ions (OH^-), carbonate ions (CO_3^{2-}) or oxide ions (O^{2-}). These ions react with acids in a number of useful reactions around the home.

When the stomach produces excess acid, people can suffer from heartburn or indigestion. Antacid tablets contain a base to neutralise the excess stomach acid.

Indicators

Indicators are chemical substances that change colour when added to acidic or basic solutions. Figure 5.5.3 shows universal indicator and Figure 5.5.4 shows bromothymol blue. Both are indicators you might use in a school laboratory. The indicators show how acidic or basic a solution is by their colour.

▲ **FIGURE 5.5.2**   In the home, bases are often used for cooking or cleaning.

base
a substance that can produce hydroxide, oxide or carbonate ions in solution

indicator
a chemical that changes colour in acidic and basic solutions

▲ **FIGURE 5.5.3**   Universal indicator is red, pink/orange or yellow in acids. It turns green in a neutral solution and blue or purple in a basic solution.

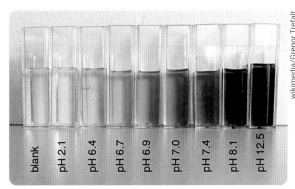

◀ **FIGURE 5.5.4**   Bromothymol blue is a simpler indicator than universal indicator. It turns yellow in acids, blue in bases and green in a neutral solution.

blank | pH 2.1 | pH 6.4 | pH 6.7 | pH 6.9 | pH 7.0 | pH 7.4 | pH 8.1 | pH 12.5

The pH scale

pH
a measure of how acidic
or basic a substance is

The **pH** scale shows how acidic or basic a substance is. Most substances have a pH between 0 and 14.

- A substance with a pH less than 7 (pH < 7) is acidic. Substances that are more acidic have a pH closer to 0.
- A substance with a pH of 7 (pH = 7) is neutral. It is neither an acid nor a base.
- A substance with a pH of greater than 7 (pH > 7) is basic. Substances that are more basic have a pH closer to 14.

The pH scale is not a regular scale, where 2 is just 1 more than 1. Instead, it is a logarithmic scale. This means that each number is 10 times more than the previous number (Figure 5.5.5).

- pH 1 is 10 times more acidic than pH 2.
- pH 3 is 10 times less acidic than pH 2.
- pH 13 is 10 times more basic than pH 12.
- pH 10 is 10 times less basic than pH 11.

Concentration of hydrogen ions compared to distilled water	pH	Examples of solutions at this pH
10 000 000	0	Battery acid, strong hydrofluoric acid
1 000 000	1	Hydrochloric acid secreted by stomach lining
100 000	2	Lemon juice, gastric acid, vinegar
10 000	3	Grapefruit juice, orange juice, soda water
1000	4	Tomato juice, acid rain
100	5	Soft drinking water, black coffee
10	6	Urine, saliva
1	7	'Pure' water
$\frac{1}{10}$	8	Seawater
$\frac{1}{100}$	9	Baking soda
$\frac{1}{1000}$	10	Milk of magnesia
$\frac{1}{10\,000}$	11	Ammonia solution
$\frac{1}{100\,000}$	12	Soapy water
$\frac{1}{1\,000\,000}$	13	Bleach, oven cleaner
$\frac{1}{10\,000\,000}$	14	Liquid drain cleaner

Increasing acidity · Neutral · Increasing basicity

▶ **FIGURE 5.5.5** The pH scale. Each level is different from the next by a factor of 10.

9780170472838

Identifying acids and bases by using indicators

You will identify the colours of indicators in acidic and basic solutions. You will then use this knowledge to classify common household substances as acids or bases.

What to do

Part 1: What colour are indicators?

- Set up three small test tubes with approximately 1 mL of 0.1 mol L^{-1} hydrochloric acid solution (test tube 1), distilled water (test tube 2) and 0.1 mol L^{-1} sodium hydroxide solution (test tube 3).
- Add 2–3 drops of universal indicator to each test tube and record the colour of the indicator in acidic, neutral and basic solutions.
- Extension: Your teacher may ask you to prepare your own indicator from red cabbage. Chop the cabbage, add boiling water from a kettle then stir the mixture for 10 minutes until the water is a purple colour. Strain the cabbage and keep the liquid to use as an indicator.

Part 2: Common substances – acid, base or neutral?

- Gather a selection of household chemicals or drinks, such as juice, soft drink, bleach, shampoo, milk or dishwashing liquid (anything you can find).
- Place approximately 1 mL of each substance into their own test tube.
- Add 2–3 drops of universal indicator to each test tube.
- Repeat with the other indicators and fresh 1 mL samples.
- Record all colours observed.

What do you think?

1 **Create** a results table showing the colours of each indicator in acid, base and neutral solutions.

2 Use this table to **identify** whether each household substance is acid, base or neutral. **Create** a summary table to show your results.

3 Were any of the household chemical results hard to determine?

4 **Describe** two difficulties you encountered while conducting this experiment that may affect the confidence you have in your results.

5 **List** any patterns in your household substances results. Do all the foods have the same acid/base nature? Do all the cleaning materials have the same acid/base nature?

---

1 The following acids are commonly found in nature and/or synthesised for human use.

- Citric acid
- Nucleic acid
- Ethanoic acid
- Amino acid
- Ascorbic acid
- Uric acid
- Folic acid

Choose three acids from the list and **research** their formulas and uses.

2 **Describe** the properties of acids and bases.

3 **Explain** what information is shown by the pH scale.

4 Phenolphthalein is colourless in an acid and pink in a base. What colour would you expect phenolphthalein to turn when added to:

a juice from a lemon?

b bleach from your laundry?

Extra science investigations
Neutralisation reactions

Making sherbet

5.6 Chemical reactions in our lives: corrosion

BY THE END OF THIS MODULE, YOU WILL BE ABLE TO:

✓ justify the conditions required for corrosion to occur
✓ explain factors that affect the rate of corrosion.

Video activity
Oxidation reactions

GET THINKING

The terms 'corrosion' and 'rusting' are often used interchangeably. This means that either word can be used to describe the same thing. In science, there are often pairs of words like this where one word is a more scientific term (corrosion), and the other word is more commonly used in everyday language (rusting). Make a list of as many of these pairs of words you can think of from any area of science.

What is corrosion?

You have probably seen a rusty nail like the one in Figure 5.6.1. The rust is the result of the process of **corrosion**, which you might know better as rusting, the most common form of corrosion. Any object made from iron can rust, such as fencing, roofing and bridges.

corrosion
a chemical process that often happens to metals exposed to oxygen; also known as rusting

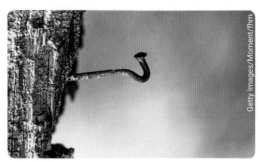

▲ **FIGURE 5.6.1** Rusting occurs when iron reacts with oxygen in the presence of water.

Corrosion is the **oxidation** of a metal. This means that oxygen is added to a metal to form a **metal oxide**. This can be seen in the general word equation:

$$\text{metal} + \text{oxygen} \rightarrow \text{metal oxide}$$

oxidation
a chemical process in which oxygen is added to a substance

metal oxide
a chemical substance made up of a metal and oxygen

The most common example you would have encountered is the rusting of iron:

$$\text{iron} + \text{oxygen} \rightarrow \text{iron oxide}$$

$$4Fe(s) + 3O_2(g) \rightarrow 2Fe_2O_3(s)$$

When iron corrodes, the iron oxide product can weaken structures, furniture and other items made from iron.

Other metals corrode, but not always as obviously as iron. You may have used magnesium ribbon in experiments in your science studies. Figure 5.6.2 shows magnesium ribbon uncleaned (left) and cleaned (right). Magnesium ribbon reacts with oxygen in the air to form magnesium oxide. This is seen as a darker layer on the surface of the magnesium metal. You should clean metals before use in experiments. This is to make sure they don't have any metal oxide on their surface because this might change the results of an experiment.

▲ FIGURE 5.6.2   Corrosion of magnesium ribbon leaves a dark layer on the surface of the metal.

Factors affecting corrosion

In most cases, metals (e.g. magnesium) react directly with oxygen in the air to form the metal oxide. However, some metals undergo a more complex corrosion process. For example, iron corrodes in a series of steps involving both oxygen and water. Without water being present, iron will not corrode very easily, if at all. This is why you often see rusted iron structures around water. Iron furniture and structures used outdoors are often coated with plastic or another protective material such as paint. This stops the iron from coming in contact with water and oxygen, and prevents corrosion.

Exposure to salt water speeds up the rate of corrosion of iron. Salt water contains sodium chloride as charged particles (ions). These ions speed up the rusting process. Saltwater exposure is a leading cause of the rusting of metals.

▲ FIGURE 5.6.3   Corrosion is a problem in mine sites and gas refineries across Australia. The Karratha gas plant in Western Australia is routinely checked for corrosion problems during shutdown inspections.

Factors that affect the rusting of iron

Set up the experiment shown in Figure 5.6.4. This works best if you use ungalvanised nails. Each test tube has a different set of conditions. Oil is used to keep oxygen out of the water. Anhydrous calcium chloride absorbs water from the air, so the environment is dry. Leave the test tubes overnight if possible.

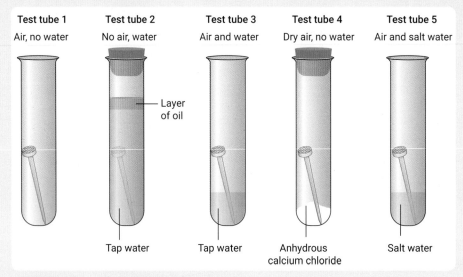

| Test tube 1 | Test tube 2 | Test tube 3 | Test tube 4 | Test tube 5 |
| Air, no water | No air, water | Air and water | Dry air, no water | Air and salt water |

▲ FIGURE 5.6.4   The experimental set-up

1   **Copy and complete** the following table by circling the conditions (oxygen, water, salt) that were present in each test tube.

Test tube	Conditions (circle those that were present)	Ranking of corrosion level (1 = least, 5 = most)
1	oxygen water salt	
2	oxygen water salt	
3	oxygen water salt	
4	oxygen water salt	
5	oxygen water salt	

2   In the final column of the table, **rank** the corrosion level of the nails. Because there are five test tubes, you should use 1–5, where 1 = least corroded, 5 = most corroded.

3   Use your results to **state** a conclusion about the conditions required for rusting of iron to occur.

4   Use your results to **state** a conclusion about the effect of salt water on corrosion.

5.6 **LEARNING CHECK**

1   When you use metals in chemical experiments, why are you often instructed to use sandpaper to clean them before use?

2   Many outdoor adventurers take their four-wheel drive vehicles and metal boats through salt water. Cars and boats that contact salt water should be washed thoroughly at the end of the day to prevent corrosion. **Explain** why this precaution is necessary.

9780170472838

5.7 Chemical reactions in our lives: reactions of bioethanol

BY THE END OF THIS MODULE, YOU WILL BE ABLE TO:

✓ describe chemical reactions involved with the production and use of bioethanol
✓ write word equations for the processes of photosynthesis, fermentation and combustion of ethanol.

GET THINKING

Think about the word 'bioethanol'. You have probably heard of ethanol and are familiar with the prefix 'bio' from earlier science studies. Why do you think they are combined? How might bioethanol be different from other types of ethanol?

bioethanol
ethanol produced from crops through the process of fermentation

ethanol
a chemical containing carbon, hydrogen and oxygen that can be used as a fuel

What is bioethanol?

Bioethanol is ethanol that is produced from plant sugars. Bioethanol is used primarily for fuel or in alcoholic drinks. **Ethanol** is also used in cleaning and in the production of plastics. Ethanol for this purpose is produced from a molecule called ethene, found in fossil fuels (Figure 5.7.1).

How is bioethanol made?

Bioethanol is made from plant crops such as corn, wheat and sugar cane. The plants produce sugars by **photosynthesis**. The sugars are then converted to ethanol by **fermentation**. This ethanol is then burned in fuels in a process called **combustion**.

Photosynthesis, fermentation and combustion are all types of chemical reactions.

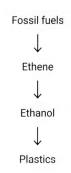

▲ **FIGURE 5.7.1**   The production of bioethanol and synthetic ethanol

Photosynthesis in plants

You may recall that photosynthesis is a chemical reaction that occurs in plants and converts carbon dioxide (CO_2) and water (H_2O) into sugar ($C_6H_{12}O_6$) and oxygen (O_2). Photosynthesis also requires light energy from the Sun (Figure 5.7.2). This makes the reaction endothermic because it absorbs energy during the reaction.

The word and chemical equations for photosynthesis are:

$$\text{carbon dioxide} + \text{water} \rightarrow \text{sugar} + \text{oxygen}$$

$$6CO_2(g) + 6H_2O(l) \rightarrow C_6H_{12}O_6(aq) + 6O_2(g)$$

The sugar that forms is called glucose. Plants can convert glucose into sucrose, which is the sugar that makes fruit taste sweet. Glucose can also be converted into starch in vegetables and crops such as potatoes, corn and wheat.

photosynthesis
the process in which producers use light energy to convert carbon dioxide and water into sugars

fermentation
the chemical reaction that uses yeast to convert glucose to ethanol

combustion
the burning of substances, such as fossil fuels, in oxygen to produce heat and light

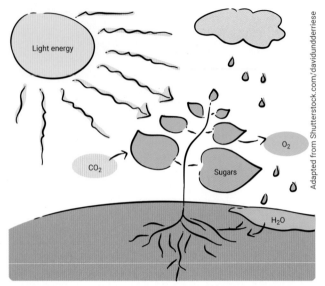

▲ FIGURE 5.7.2 The process of photosynthesis in plants

Fermentation of plant sugars

Fermentation is the process of converting glucose to ethanol, the glucose being sourced from crops such as wheat, corn and sugar cane. All these plants contain starch or glucose that can be fermented.

The word and chemical equations for fermentation are:

$$\text{glucose} \rightarrow \text{ethanol} + \text{carbon dioxide}$$

$$C_6H_{12}O_6(aq) \xrightarrow[\text{heat}]{\text{yeast}} 2C_2H_5OH(aq) + 2CO_2(g)$$

Fermentation requires a specific set of conditions. The mixture needs to be warm, slightly acidic and have no oxygen present. Yeast is required.

Yeast is a type of micro-organism used to make bread and alcohol in beer and wine. The yeast converts sugar into ethanol. Yeast is sensitive to acid levels and temperature. If the conditions are not right, the yeast will die. For example, if oxygen is present, then ethanol will not form, and the resulting liquid will taste like vinegar.

After fermentation, the final mixture is processed and purified, and the resulting ethanol is used to make alcoholic drinks or fuel for cars (Figure 5.7.3). In Module 5.9, you will learn more about the use of ethanol as a fuel for cars.

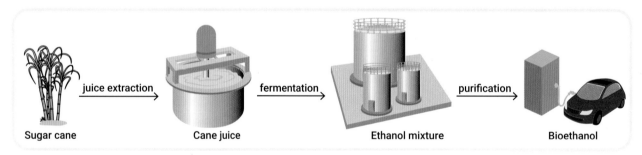

▲ FIGURE 5.7.3 The fermentation of plant sugar to bioethanol

Combustion of fuels

Cars can burn ethanol as fuel in the presence of oxygen. This process is called combustion. Combustion is also the process that occurs when we burn fossil fuels such as petrol or barbecue gas. Wood and coal also combust in fireplaces, campfires and barbecues.

Combustion is the rapid reaction of a fuel with oxygen to produce energy in the form of heat and light. Most substances we combust are carbon-based fuels such as petrol, diesel, bioethanol, wood and coal. When carbon-based fuels are combusted, they can undergo **complete combustion** or **incomplete combustion**.

complete combustion combustion occurring in excess oxygen, producing carbon dioxide and water

incomplete combustion combustion occurring in limited oxygen, producing carbon, carbon monoxide and water

9780170472838

Complete combustion

The products of complete combustion of a carbon-based fuel are carbon dioxide (CO_2) and water. This is represented by the word equation:

$$\text{fuel} + \text{oxygen} \rightarrow \text{carbon dioxide} + \text{water}$$

Incomplete combustion

During incomplete combustion, the available oxygen is limited. As the fuel burns, there is not enough oxygen to completely combust the fuel.

The products of incomplete combustion have less oxygen than the products of complete combustion. The usual products are either carbon monoxide (CO) or carbon (C), or a mixture of both. Water is also produced.

Incomplete combustion can be represented as:

$$\text{fuel} + \text{oxygen} \rightarrow \text{carbon monoxide} + \text{water}$$

$$\text{fuel} + \text{oxygen} \rightarrow \text{carbon monoxide} + \text{carbon} + \text{water}$$

Complete and incomplete combustion can be observed with a Bunsen burner (you will investigate this in the activity below).

Observations of combustion

You can see incomplete and complete combustion by looking at the colour of a flame and the emissions produced (Figure 5.7.4).

Another sign of incomplete combustion is black particles being emitted from the flame. These particles are commonly called 'soot' and are particles of carbon (C).

▲ FIGURE 5.7.4 (a) Complete and (b) incomplete combustion can be identified by the colour of the flame.

Complete and incomplete combustion in a Bunsen burner

☆ ACTIVITY

Carefully light a Bunsen burner and observe the flame with the airhole fully opened (maximum oxygen). Then observe the flame with the airhole closed (reduced oxygen).

1 **Compare** the two flames. Record the differences you observe between the two flames.
2 **Identify** which flame is showing complete combustion. Use your observations to **justify** your selection.
3 Bunsen burner gas is rich in methane (CH_4). **Write** a balanced equation for the complete combustion of methane.

 **LEARNING CHECK**

1 **Write** word and chemical equations for the processes of:
   a photosynthesis.
   b fermentation.
   c the complete combustion of butane (C_4H_{10}) (used in gas bottles in camping stoves).
2 **Describe** the conditions required for the fermentation of sugars to produce ethanol.

Video activities
Biofuels

Oxygen and combustion

Interactive resource
Crossword: Chemical reactions

5.8 First Nations Australians' chemistry to develop pigments and dyes

IN THIS MODULE, YOU WILL:

✓ investigate the ways First Nations Australians developed pigments and dyes through understanding of chemical reactions.

What is in paint?

First Nations Australians use paint to record and communicate knowledge. We can see this in rock and bark paintings, body decorations and features added to implements. Paints are a liquid mixture containing pigments or dyes that leave a solid film on drying. Binders can be added to thicken paints and fixatives added to provide durability once paint is applied to a surface. Many traditional paints used by First Nations Australians are prepared using chemical processes.

Calcination and pyrolysis

Calcination is the controlled thermal treatment of a solid compound. Calcination involves removing water, carbon dioxide and sulfur dioxide, and oxidising the substance through controlled heating. By calcination, some First Nations Australians produce a paint pigment of yellow ochre (limonite). The yellow limonite is carefully heat treated to chemically convert the pigment into haematite, a red-brown pigment.

Pyrolysis is the decomposition of organic matter at high temperatures and with limited oxygen that changes the chemical composition of a material. Many First Nations Australians produce charcoal by pyrolysis to make a pigment for paint. Particular woods are selected and exposed to high heat with limited oxygen to produce charcoals. Heating causes the water in the wood to evaporate before the wood is converted into carbon. The processes release compounds with a high vapour pressure, including combustible gases such as carbon monoxide, hydrogen and methane. This leaves a carbon-rich solid residue, charcoal. The charcoal is collected, ground into a powder and combined with other products to make paint.

▲ **FIGURE 5.8.1** Rock paintings in the Cathedral Caves in Carnarvon Gorge, Queensland, on the lands of the Bidjara and Karingbal Peoples

Some examples of paints

Mineral pigments

Table 5.8.1 gives some examples of the different types of mineral pigments used by First Nations Australians.

▲ **FIGURE 5.8.2**   Making red ochre pigment

▼ **TABLE 5.8.1**   Some mineral pigments used by First Nations Australians

Name and chemical component	Details
White huntite, $Mg_3Ca(CO_3)_4$	• Used by the Ngarinyin Peoples of the north-west Kimberley region (Western Australia) in rock painting • The pigment is powdery and can flake from surfaces • Sites were regularly revisited to repaint illustrations
Yellow ochre (limonite), $FeO(OH)$ Red ochre (haematite), Fe_2O_3	• Used by Tasmanian First Nations Peoples and mixed with animal fat, blood, saliva or water • Is more resistant to degradation than some other mineral pigments
Pyrolusite, MnO_2	• Used in rock paintings of the Cathedral Caves in the Carnarvon Gorge by the Bidjara and Karingbal Peoples

Plant pigments

Some First Nations Australians use the pigments from plants to produce paints that are used for decorating implements or as body paint, as shown in Table 5.8.2.

◀ **FIGURE 5.8.3** The bright red fruits of the saltbush are used to make face paints by some First Nations Australians.

▼ TABLE 5.8.2 Some plant pigments used by First Nations Australians

Name	Details
Red fruits of saltbush	• Used to make skin paint, e.g. by Arrernte Peoples (Northern Territory) and the Wurundjeri Peoples (Victoria)
Bark of mountain ash tree	• Used to produce a red-brown paint for decorating implements such as spears, e.g. by the Anguthimri People (north Queensland)

Binders

Binders are chemicals used to stick materials or objects together. First Nations Australians use a variety of substances as binders, as shown in Table 5.8.3.

▼ TABLE 5.8.3 Some binders used by First Nations Australians

Name	Details
Sap of the native orchid	• A sticky and viscous substance used as a binder, e.g. by the Anindilyakwa Peoples of Groote Eylandt (Northern Territory)
Wax or honey of native bees Turtle egg yolk	• Used as binders to reduce paint flaking from surfaces, e.g. by Tiwi Peoples (Northern Territory)
Animal fats	• Used by the Barngarla Peoples (South Australia) mixed with charcoal as body paint, which makes the paint less soluble in water

Fixatives

Fixatives are substances used to fix other chemicals, such as paints, so they last longer. Some examples of fixatives used by First Nations Australians are shown in Table 5.8.4.

▲ **FIGURE 5.8.4** Nuts of the candlenut tree are used to make oil, which is used as a fixative.

▼ TABLE 5.8.4 Some fixatives used by First Nations Australians

Name	Details
Emu fat	• Used by the Ngaatjatjarra People (Western Australia) as a hydrophobic fixative for rock painting
Resinous material from yellow tea tree	• Used by the Anguthimri People (north Queensland), who mixed the resin with pigment, warmed the paint and applied it to permanently paint implements
Candlenut oil	• Used by the Walmbaria Peoples (north Queensland) to fix paints to implements

9780170472838

Making paints

You need
- ochre or ochre powder
- charcoal
- water
- egg yolk
- lard or oil
- water in a spray bottle
- mortar and pestle
- 3 stones
- paintbrushes
- 3 plastic containers

What to do

1. Choose either ochre or charcoal as your pigment.
2. Grind the ochre or charcoal in the mortar and pestle until it is a fine powder.
3. Label three containers 'Egg yolk', 'Water' and 'Fat/oil'.
4. Add a teaspoon of the powder to each of three containers.
5. Add 5 mL of water to each container and mix.
6. Add an egg yolk to the container labelled 'Egg yolk'. Mix.
7. Add 5 mL of oil or a tablespoon of lard to the container labelled 'Fat/oil'. Mix.
8. Add 5 mL of water to the container labelled 'Water'. Mix.
9. Paint the entire surface of each of the stones with one of the mixtures. Make sure you label the stones.
10. Allow the paint to dry.
11. Spray the stones with a fine water mist. Record your immediate observations. What happened to each of the mixtures? What does the mist of water represent?
12. Place the stones outside.
13. Make a table for observations and record what you notice every day for a fortnight.
14. Look for changes to the paint, such as flaking, fading, cracking or powdering.
15. Record observations about the weather.

What do you think?

At the end of the fortnight, consider your results.

1. Which paint stayed the brightest?
2. Which paint lasted the longest?
3. How did the weather affect each of the paints?
4. What other non-toxic natural paint components could you use from your local environment?
5. What does this experiment help you understand about First Nations Australians' achievements in producing long-lasting paint?

5.9 Bioethanol as a more environmentally friendly fuel

BY THE END OF THIS MODULE, YOU WILL BE ABLE TO:
- ✓ explain the need for the development of bioethanol
- ✓ explain why bioethanol is considered to be more environmentally friendly than fossil fuels.

Video activity
Algae as a biofuel

Why do we need environmentally friendly fuels?

In Chapter 6, you will look at the issue of climate change due to the enhanced greenhouse effect. This is a result of the release of heat-trapping gases into the air, such as carbon dioxide from fuel combustion. Scientists are racing to develop environmentally friendlier fuels to try to combat climate change.

Why is bioethanol considered more environmentally friendly than fossil fuels?

As discussed in Module 5.7, bioethanol is produced from crops and added to petrol. In Australia, this is sold in many places as E10 fuel. This means it contains about 90 per cent regular petrol and 10 per cent ethanol. Ethanol burns 'cleaner' than regular petrol. This means fewer pollutants such as tiny particles, soot and carbon monoxide form.

When fossil fuels are combusted, they produce carbon dioxide (as discussed in Module 5.7). Carbon dioxide was originally taken out of the atmosphere by plants millions of years ago when fossil fuels formed from plant and animal remains. This means we add *extra* carbon dioxide to the atmosphere when we burn fossil fuels. As you will learn in Chapter 6, extra carbon dioxide in the atmosphere is a major contributor to climate change.

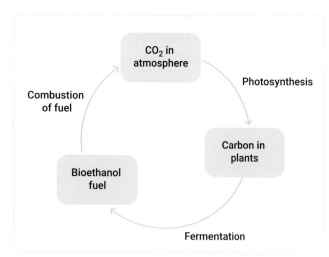

▲ **FIGURE 5.9.1** The cycle of carbon in bioethanol. Carbon dioxide from the atmosphere is taken up by plants during photosynthesis. The carbon moves from glucose to ethanol in fermentation, then back to carbon dioxide during combustion.

At first glance, it seems bioethanol isn't any different. When manufactured and combusted, it still produces carbon dioxide. But the crops used in bioethanol production only recently took the carbon dioxide out of the atmosphere by photosynthesis, and the crops are replaced with more plants. This means the carbon dioxide produced during combustion of bioethanol is just replacing and not adding to the carbon dioxide in the atmosphere, as shown in Figure 5.9.1. When combusted, biofuels also often produce slightly less carbon dioxide than regular petrol.

Bioethanol in vehicle fuel is now used in many countries and is available at many Australian service stations.

What other factors affect the production of bioethanol?

When deciding whether bioethanol is a more sustainable and environmentally friendly fuel than fossil fuels or other options, there are other factors that need to be considered. Some of these factors are listed in Table 5.9.1

▼ TABLE 5.9.1   Factors to consider when assessing the production of bioethanol

Factor	What does it have to do with bioethanol?	Why is this an issue?
Land use	A large amount of land is required to grow the crops used to make bioethanol.	Land clearing for agriculture means cutting down trees that take in carbon dioxide. Fewer trees taking in carbon dioxide contributes to climate change. It also means that less land is available for growing food crops or grazing animals. Overall, more land is used for agriculture (food) and bioethanol than for regular agriculture alone.
Use of crops as fuel	The crops used for bioethanol cannot be used as food.	A large part of the world population currently does not have access to enough food. Many countries cannot grow enough crops to provide food for their people because of droughts, floods, fire, war and other recent world events. Some people think it is unethical to use food as fuel when so many people cannot get enough food to eat.
Industrial production	The production of bioethanol requires large amounts of energy for the different processing stages.	The energy for the conversion of crops to bioethanol requires several processes that need either fuel or electricity. Most fuels are fossil fuels such as gas or coal. Most electricity comes from fossil-fuel-powered power stations. This releases carbon dioxide into the atmosphere extra to that produced during combustion.
Transport	The crops need to be transported to the processing facility. The bioethanol is then transported around the world and across countries where it is sold.	Most transport is by trucks or trains around Australia, or by ships overseas. These vehicles usually run on fossil fuels, adding carbon dioxide to the atmosphere.

Class debate

☆ ACTIVITY

Split into two teams and choose speakers to represent your team. Discuss with your teacher how many speakers will make up each team. Depending on the number of speakers, speeches could be 2–5 minutes long.

The question for debate is: 'Should Australia produce and use more bioethanol?'

Your team needs to come up with arguments for or against (depending on your side). You should use the ideas in Table 5.9.1 and do some extra research to find facts and examples that support your arguments.

1 A scientist produced the graph in Figure 5.9.2 after conducting a series of experiments to compare petrol and E10 as fuels, to provide drivers with more information about both fuels so they can make an informed choice. The graph shows the fuel consumption (litres of fuel used to travel 100 km) of petrol and E10 (ethanol + petrol).

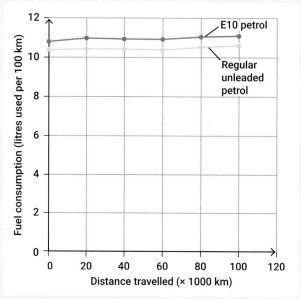

▲ **FIGURE 5.9.2** A comparison of petrol and E10 fuel consumption

    a What does the 10 mean in E10 fuel?

    b Write a conclusion about the difference in fuel consumption of the two fuels based on the graph.

2 Petrol and bioethanol both release carbon dioxide into the atmosphere. **Explain** why bioethanol is considered to be better environmentally than fossil fuels.

Shutterstock.com/lstimages

▲ **FIGURE 5.9.3** Many petrol stations in Australia sell E10 fuel.

3 In pairs, groups or individually, **create** a poster or advertisement that explains why E10 is a more environmentally friendly fuel than petrol.

4 **Research** to find out why 100 per cent ethanol cannot be sold as car fuel. **Write** a few sentences to summarise what you found.

Controlling variables to perform a valid investigation

SCIENCE SKILLS IN FOCUS

IN THIS MODULE, YOU WILL FOCUS ON LEARNING AND IMPROVING THESE SKILLS:

▶ designing and assessing controlled investigations, including control of variables and use of a control

▶ assessing the validity and reproducibility of methods.

CONTROLLING VARIABLES

Controlled variables are important for ensuring the validity of an investigation. What do we mean by this? A valid investigation is one in which one variable is changed (the independent variable) to determine the effect on a second variable (the dependent variable).

An example of this is an investigation to determine how the mass of a chemical reactant affects the amount of product formed. In this case, the independent variable is the mass of the reactant. The dependent variable is the amount of product formed.

For this investigation to be valid, only *one* variable (the independent variable) must be changed. For example, if the temperature of the mixture also changed, how can you be sure it was the mass of the reactant that caused the change in the amount of product? How can you be sure the temperature had no effect? If they both had an effect on the dependent variable, how do you know how much each contributed?

When designing an experiment, it is important to not only identify controlled variables, but describe *how* you will keep them controlled. For example, students often write 'the volume of water will be controlled'. A better description would be '100 mL of water will be used in each trial. This will be measured with a measuring cylinder, so it is always the same volume'.

When you perform the experiments in this module, focus on trying to describe controlling variables in detail and include *how* you will control them.

A control is useful in some experiments. A control is a trial with no independent variable present. It is part of the method and is used as a comparison to find out if the independent variable had an effect on the dependent variable. It helps us find out if an effect can be attributed to one cause. In the previous example, you would perform the experiment with no reactant to see if the product would form. There are two possible outcomes:

▶ Outcome 1: no product forms suggesting that there is no other factor that causes the product to form.

▶ Outcome 2: some product forms suggesting that there is some other factor (e.g. temperature) that is causing the product to form without reactant. (This should strike you as odd, and it is. With no reactant, you should not expect any product to form.)

A control is not always appropriate. In the next two experiments, there are questions about controls. In one experiment, a control is appropriate. In the other, it is not. When considering whether a control is appropriate, try to predict whether anything is likely to occur if the independent variable is not changed. In chemistry, if you remove the independent variable and no chemical reaction occurs, then a control is not appropriate.

Science skills resource
Science skills in practice: Controlling variables for a valid investigation

Video
Science skills in a minute: Controlling variables

INVESTIGATION 1: HOW DOES THE AMOUNT OF SUGAR AFFECT THE RATE OF PRODUCTION OF ETHANOL?

AIM

To design a controlled investigation to determine the effect of increasing the amount of sugar present on the rate of production of ethanol

MATERIALS

This investigation will be done as a class because each trial can take some time. For each trial, each group will need:

☑ 250 mL conical flask
☑ electronic balance (reading to 0.01 g)
☑ 5 g yeast (e.g. baker's yeast)
☑ a mass of glucose, sucrose or table sugar
☑ 40 mL warm water (30–40°C)
☑ airlock – you can make a simple airlock with a piece of bendable hose, a rubber band and some water – see Figure 5.10.1. Once you have made a circle with the hose, use a rubber band to keep it in the shape shown. Then pour some water into the hose so it sits at the bottom of the circle part of the airlock. The water will keep oxygen out but allow any gases from the experiment to escape.

▲ **FIGURE 5.10.1** A simple airlock

METHOD

The following method is for a single trial.

1 Add 5 g of yeast, 10 g of glucose and 40 mL of warm water to the flask, add the airlock and swirl to mix the contents.

2 Immediately weigh the flask and its contents and record the mass.

3 After a set period of time (see class design instructions below), reweigh the flask and record the mass.

4 The mass difference is the amount of carbon dioxide lost to the atmosphere in this chemical reaction.

As a class, you are going to design an experiment using steps 1–3 to see how the mass loss changes as you change the amount of glucose used. You need to consider the following points when designing your method to ensure the experiment is fair and controlled.

- If you are adjusting the level of glucose used, what values are you going to use as the different levels of glucose?
- What other quantities will need to be controlled (kept the same) in each trial?
- How many trials are suitable for the same level of glucose?
- Are you going to set up a control? How would you do this?
- How long are you going to leave the mixture to ferment? Your teacher may be able to store your conical flasks overnight; otherwise, leave them as long as you can during the lesson for the best results.
- How are you going to measure the dependent variable? What units will be used?

RESULTS

1 Create a table of results showing the mass of glucose used and the mass loss from each flask. Determine what data can be averaged if you performed trials for the same mass.

2 Create a graph with mass of glucose on the horizontal axis and mass loss on the vertical axis. Consider whether a straight line of best fit or a curve would be best suited for your data.

9780170472838

EVALUATION

1 The mass loss in the conical flask is due to the formation and escape of carbon dioxide during the fermentation process. Write a word (or balanced) equation to explain how the mass loss from each conical flask indicates how much ethanol has formed in each flask.

2 Explain the importance of controlling the quantities of substances in each trial of an experiment. Use specific examples from this experiment in your answer.

3 Explain how a control could be used in this experiment. If you used one in your class experiment, describe how it was set up.

4 Describe the steps you took in your class experimental design to ensure the validity of the experiment.

5 Explain why it is good practice to perform trials for each level of the glucose in this experiment.

CONCLUSION

Review the aims and results of your investigation. Write a sentence explaining whether you achieved the aim of the investigation.

INVESTIGATION 2: DO ALL FUELS PRODUCE THE SAME AMOUNT OF ENERGY?

AIM

To compare how different fuels produce energy

Warning

- Using spirit burners will protect you from the fuels, which are highly flammable. Use gloves to handle the spirit burners and do not touch the wicks or you will come into contact with the fuel.

- Be careful with the open flame; keep all paper and clothes at a safe distance. Extinguish flames with the spirit burner lid, not by blowing them out.

MATERIALS

- ☑ spirit burners with three different fuels. Your teacher will let you know what fuels you have in your experiment. Try to use ethanol as one of your fuels. Other possible fuels are methanol and kerosene.
- ☑ 250 mL conical flask
- ☑ 100 mL measuring cylinder
- ☑ thermometer or data logger temperature probe
- ☑ clamps and retort stand
- ☑ matches
- ☑ heatproof mat
- ☑ stopwatch (or phone with timer)

METHOD

1 Set up equipment as shown in Figure 5.10.2. Keep the conical flask at a fixed height above the spirit burner.

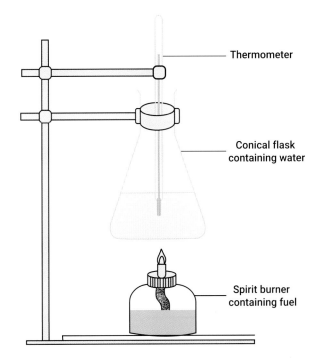

▲ FIGURE 5.10.2   The equipment set-up

2. Measure 100 mL of water with the measuring cylinder and place into the conical flask.

3. Insert the thermometer and record the initial temperature of the water.

4. Light the spirit burner, place it under the conical flask, and heat the water for 1 minute. After 1 minute, cap the spirit burner and record the final temperature of the water.

5. Repeat steps 2–4 with the other two fuels.

6. Repeat steps 2–5 twice more so you do three trials of each fuel.

RESULTS

Copy Table 5.10.1 into your workbook to record your results data. You will need to create one table for each fuel. If you have Excel, use a spreadsheet to record the data.

▼ TABLE 5.10.1   Results table

Trial	Name of fuel:		
	Initial temp. (°C)	Final temp. (°C)	Change in temp. (°C)
1			
2			
3			

EVALUATION

1. Describe the steps taken in the method to ensure that:
   a. a valid test was conducted – this means the experiment tested the aim of the experiment
   b. reliable results were produced – this means that each time you did a trial for the same fuel, you should have measured almost the same result.

2. Explain why a control is not suitable for this investigation.

3. Suggest why there was variation between the:
   a. different fuels. (Why was the average temperature change different?)
   b. trials of the same fuel. (Why was the data not identical in trials 1, 2 and 3?)

CONCLUSION

Review the aim and hypothesis and write a conclusion to this investigation. Refer to specific result data in your conclusion.

REMEMBERING

1 A chemical reaction always has chemical change occurring. What observations might indicate a chemical reaction has occurred?

2 **State** the law of conservation of mass.

3 a What information is provided by a chemical word equation?

   b What extra information (in addition to what you identified in part a) is provided by a balanced formula equation?

4 **Explain** why it is not possible to just change the chemical formula to make an equation balance. Use an example in your answer.

5 **State** two chemical processes that are used to produce bioethanol.

6 **Describe** the conditions required for corrosion of iron to occur.

7 **Identify** two substances in or around your house that are acidic, and two substances that are basic.

UNDERSTANDING

8 When 16 g of methane is added to oxygen, 44 g of carbon dioxide and 36 g of water is formed. What is the mass of oxygen used in this reaction?

9 Use the law of conservation of mass to **explain** why formula equations need to be balanced.

10 **Explain** why millions of dollars are spent each year protecting structures such as bridges and jetties from corrosion.

11 **Explain** the difference between complete and incomplete combustion.

12 **Explain** the process of pyrolysis and how this relates to the paints produced by many First Nations Australians.

APPLYING

13 **Balance** the following equations.

   a $Cl_2 + NaBr \rightarrow NaCl + Br_2$

   b $CuCl_2 + H_2S \rightarrow CuS + HCl$

   c $C_5H_{12} + O_2 \rightarrow CO_2 + H_2O$

   d $Ca + HCl \rightarrow CaCl_2 + H_2$

   e $BaCO_3 + HCl \rightarrow BaCl_2 + CO_2 + H_2O$

   f $P_4O_{10} + H_2O \rightarrow H_3PO_4$

14 The table shows several indicators and their colours in acidic and basic solutions.

Indicator	Colour in acid	Colour in base
Methyl orange	Red	Yellow
Bromophenol blue	Yellow	Blue
Thymol blue	Yellow	Blue
Phenolphthalein	Colourless	Pink
Alizarin yellow	Yellow	Red

   a What colour would you observe if hydrochloric acid was added to methyl orange?

   b A student was using methyl orange, thymol blue, alizarin yellow and a range of common household substances. They formed a conclusion that the substances tested were all acids because they all changed the colour of the indicators to yellow. **Explain** why the conclusion may not be accurate.

   c **Explain** how the student could modify the experiment in part **b** to have a more accurate conclusion.

15 Cars take in air from the atmosphere to get oxygen needed for combustion. Cars have air filters that block pollutant particles from the air. These filters are replaced regularly as they become blocked. You can tell when your air filters are blocked because the exhaust starts to release black smoke. **Explain** why black smoke can tell you your air filters need replacing.

16 All exposed iron metal on the Sydney Harbour Bridge is painted every 5 years. In Brisbane, the Story Bridge gets painted every 7 years. **Explain**, in terms of corrosion, why painting prevents corrosion.

17 Race cars like the Australian Supercars use E85 fuel. What is the composition of E85 fuel?

18 In Australia, a particular brand of car has a fuel efficiency of 7.8 L per 100 km when using unleaded fuel. When using E10, the car is 15 per cent less fuel efficient. What is the fuel efficiency of the car when using E10 fuel?

19 In the following chemical reaction, the black sphere is a carbon atom, the white spheres are hydrogen atoms, and the red spheres are oxygen atoms. The CH_4 molecule is called methane.

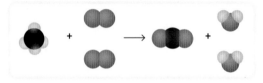

a   Is the reaction balanced? **Explain** your answer.

b   **Write** a word equation and a balanced equation for this reaction.

c   Less energy is required to break the bonds in the reactants than was released in the formation of the products. **Predict** and **explain** whether this reaction is endothermic or exothermic.

EVALUATING

20 A student conducted an experiment using household kitchen chemicals. They added 20 mL of vinegar to 5 g of bicarbonate of soda in an open container. Vinegar contains ethanoic acid and the chemical name for bicarbonate of soda is sodium hydrogen carbonate. When these chemicals reacted, bubbles formed. After the reaction was completed, the student found that 18 g of product was formed. He made the following conclusion:

In this experiment, the law of conservation of mass was shown to be wrong because the masses of reactants and products were not equal.

By considering the reactants and products in this chemical reaction, **evaluate** the student's conclusion and **state** whether or not he is correct.

21 As seen in Question 17, fuel efficiency in most cars drops by around 15 per cent when using E10 fuel compared to regular unleaded petrol. **Assess** the benefits and problems that a driver may consider when deciding to use unleaded or E10 fuel.

CREATING

22 **Create** a stop-motion animation to show what happens to the atoms as nitrogen gas and hydrogen gas form ammonia gas. The following images of the molecules should help get you started.

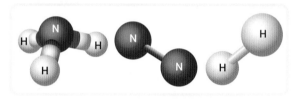

23 During the production and combustion of bioethanol, carbon takes on different forms. **Create** a flow chart that shows how carbon moves from the atmosphere, through the processes of photosynthesis, fermentation and combustion and back to the atmosphere. Include the chemical formula of the carbon substance at each step and word or balanced equations to show how the carbon changes at each stage.

24 Fermentation of glucose can only occur within a certain range of temperatures. The yeast required for fermentation die if conditions get too warm or too cold. Using Investigation 1 on page 172 as a guide, design an experiment to find the optimum temperature for fermentation of glucose.

Note that the optimum temperature is when the reaction is occurring the fastest. In your experiment, you need to think about how you will determine the fastest reaction. What will you measure? How will it tell you whether the reaction is fast or slow?

You will also need to think about controlling variables. What needs to stay constant? What needs to change?

BIG SCIENCE CHALLENGE PROJECT #5

1 Connect what you've learned

In this chapter, you have learned about chemical reactions and representing them by word and balanced equations. You have also investigated a range of chemical reactions involving bioethanol.

Create a flow chart to show the steps in the production of bioethanol from crops (glucose) to the combustion of ethanol in car engines. Show where energy is lost or required in the process.

2 Check your thinking

At the start of the chapter, you were asked about the use of bioethanol as a more environmentally friendly fuel. What is your understanding of that concept now? Write a paragraph, draw a diagram or use any method you like to explain your understanding of why burning bioethanol is better for the environment than burning fossil fuels.

You were also asked whether you could see any problems with using food and crops for fuel. Do some research, and/or have a group discussion about the pros and cons of using crops to make bioethanol instead of feeding people.

Alamy Stock Photo/All Canada Photost

3 Make an action plan

At the start of the chapter, you were asked to find out how easy it is to find E10 fuel where you live. Most states have a fuel watch website you could search to find this information. Does your area have a lot of E10 fuel available? Some states in Australia have E10 fuel nearly everywhere; in other states it is not available except at a few petrol stations. Can you find out which states have E10 available, and which do not?

Do you think E10 fuel should be available everywhere in Australia to give people a choice about the fuel they use? Justify your choice.

4 Communicate

Create an advertisement that could be shown on social media or TV that encourages people to consider using E10 fuel in their cars.

6 The carbon cycle

SOUTHERN Biological

Shutterstock.com/NadyGinzburg

FIGURE 6.0.1 Vehicles emit carbon dioxide, one of the greenhouse gases that is contributing to climate change.

Every time you use a car, train or bus, the vehicle emits pollutants. These are the products of the combustion reaction that occurs inside petrol and diesel engines. The emissions include carbon dioxide, which is a greenhouse gas. Our increased use of vehicles that rely on the combustion of fossil fuels is changing the carbon cycle.

The extra carbon dioxide we're adding to the atmosphere results in more heat being trapped. This in turn is warming Earth and changing the climate faster than we – and plants and animals – can adapt.

▶ How often do you ride in a car?

▶ Can you think of other human activities, such as food production and transport, that contribute to extra carbon in our atmosphere?

#6 SCIENCE CHALLENGE ACCEPTED!

At the end of this chapter, you can complete the Big Science Challenge Project #6. You can use the information you learn in this chapter to complete the project.

Assessments
- Prior knowledge quiz
- Chapter review questions
- End-of-chapter test
- Portfolio assessment task: Science data investigation

Videos
- Science skills in a minute: Reproducible methods and accurate results **(6.9)**
- Video activities: The water cycle **(6.1)**; Photosynthesis **(6.2)**; Oxygen and combustion **(6.3)**; The carbon cycle **(6.4)**; Climate change **(6.5)**; The greenhouse effect **(6.5)**; What is carbon farming? **(6.6)**; Cape Grim monitoring station **(6.8)**

Science skills resources
- Science skills in practice: Reproducible methods and accurate results **(6.9)**
- Extra science investigations: Greenhouse effect in a bottle **(6.5)**; Effect of CO_2 on water **(6.6)**

Interactive resources
- Simulation: Greenhouse effect **(6.5)**
- Drag and drop: Earth's spheres **(6.1)**; Natural v enhanced greenhouse effect **(6.5)**
- Label: Photosynthesis v cellular respiration **(6.2)**; Carbon cycle **(6.4)**

Nelson MindTap

To access these resources and many more, visit:
cengage.com.au/nelsonmindtap

Video activity
The water cycle

Interactive resource
Drag and drop: Earth's spheres

GET THINKING

You might remember learning about the water cycle, and how water moves through Earth's oceans, rivers, groundwater and atmosphere. How do you think the water cycle might be related to Earth's four spheres?

biosphere
all the parts of Earth and the atmosphere that support life, including all living things

hydrosphere
the component of Earth's system that consists of all the solid, liquid and gaseous forms of water

geosphere
the rocks, minerals and landforms of Earth's surface and its interior

atmosphere
all of the gases surrounding Earth's surface

Earth has four spheres

Scientists think of Earth as having four major systems that interact with each other. These are known as spheres: **biosphere**, **hydrosphere**, **geosphere** and **atmosphere** (Figure 6.1.1). Thinking of Earth as interconnected spheres helps scientists understand how energy and matter move and interact with biotic (living) and abiotic (non-living) things.

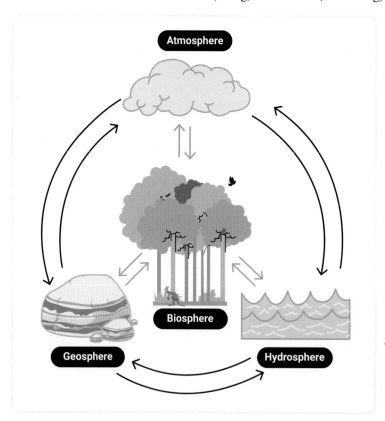

FIGURE 6.1.1 Earth's four spheres are the biosphere, the hydrosphere, the geosphere and the atmosphere.

Biosphere

The biosphere is the component of Earth where living things, such as plants, animals and bacteria, are found. Living things, also called organisms, live in oceans, under rocks and in the atmosphere. Organisms have even been found inside volcanic vents and in Earth's deepest ocean trenches – about 11 km under water! Abiotic factors, such as temperature, rainfall and access to fresh water, affect the environment and help determine where different types of organisms live.

Hydrosphere

The hydrosphere refers to the solid, liquid and water vapour forms of water on Earth. The hydrosphere includes water in oceans, lakes, rivers, snow, ice, glaciers and groundwater on or below Earth's surface. When water evaporates, it interacts with the atmosphere as water vapour. When water **condenses**, it can fall back to Earth as **precipitation**. The Sun's energy drives changes in the hydrosphere and causes water to circulate around the hydrosphere in the water **cycle**.

The water cycle consists of processes such as transpiration, condensation, precipitation, run-off, **infiltration**, **percolation** and evaporation. When water travels from plant roots up to leaves and evaporates, the process is called transpiration. Refer to Figure 6.1.2 to remind yourself of the water cycle and relate the water cycle to the hydrosphere.

condense
to change state from a gas to a liquid

precipitation
liquid or solid water in the form of rain, snow, sleet or hail

cycle
continuous, connected and repeated processes that move and transform matter through Earth's spheres

infiltration
movement of water into the surface

percolation
movement (soaking) of water into soil at the ground level of Earth

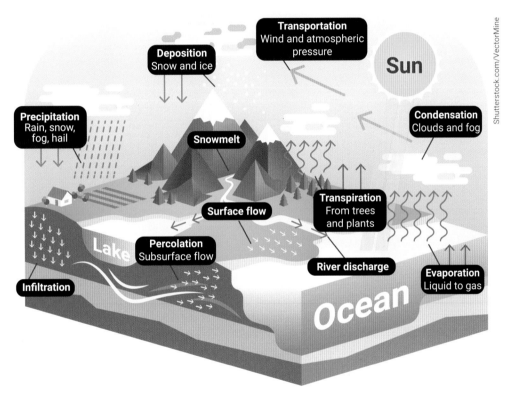

Shutterstock.com/VectorMine

◀ **FIGURE 6.1.2** The water cycle is made up of processes that circulate water around Earth.

Geosphere

The geosphere extends from Earth's surface to its core. The geosphere provides a solid foundation for living things.

The geosphere includes the rocks, minerals and landforms that make up Earth's surface, including the sediments under oceans, lakes and rivers. Volcanos, valleys, deserts, beach sand and rocks are all part of the geosphere. The geosphere also includes all the layers under Earth's surface and the rock cycle.

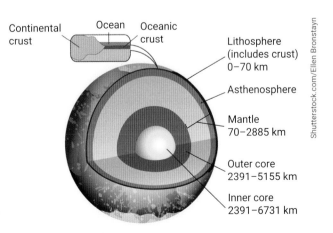

Shutterstock.com/Ellen Bronstayn

▲ **FIGURE 6.1.3** The geosphere is made up of all the layers under Earth's surface and the rocks, minerals and landforms that form Earth's surface.

Atmosphere

The atmosphere refers to gases that make up the air above Earth's surface, below outer space. The atmosphere plays a large role in storing, moving and circulating matter such as **carbon** and water, and energy such as heat. The heat stored in the atmosphere keeps Earth at an average temperature that is suitable for life.

Earth's atmosphere can be divided into five layers: troposphere, stratosphere, mesosphere, thermosphere and exosphere (Figure 6.1.4). The weather occurs in the lowest layer of the atmosphere, the troposphere.

carbon
an element in all organisms; its atoms can form four bonds that store energy

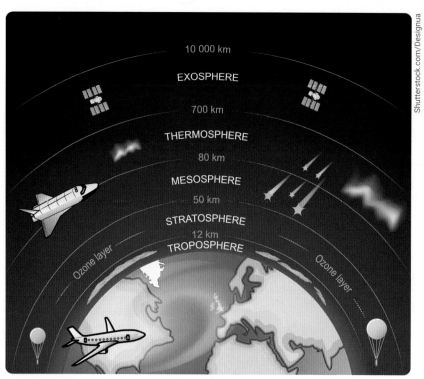

▲ **FIGURE 6.1.4**   The layers of Earth's atmosphere

As shown in Figure 6.1.5, the atmosphere is made up of 78 per cent nitrogen and 21 per cent oxygen. The remaining 1 per cent of air is made up of trace gases such as argon and carbon dioxide. The gases carbon dioxide and oxygen are essential to life, as you will see in Module 6.2.

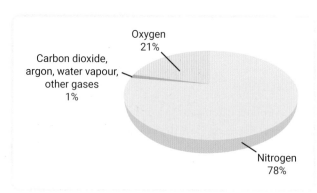

▲ **FIGURE 6.1.5**   The composition of air

9780170472838

Earth's spheres interact

Earth's four spheres interact in complex ways as one large **system**. Energy and matter circulate through the four interacting spheres through natural processes and cycles, such as the water cycle and **carbon cycle**. Matter is transformed (changes form) or transferred (changes location) in processes such as photosynthesis, cellular respiration, combustion, evaporation, ocean and atmospheric currents, weathering and erosion, and volcanic eruptions.

system
a set of interacting processes and components

carbon cycle
the transfer and transformation of carbon through Earth's spheres

 LEARNING CHECK

1 **List** the four spheres found on Earth.

2 **Define** 'cycle'.

3 **Name** a substance that circulates around Earth's spheres, and list the processes that drive its transfer into different spheres or transformation into different forms.

4 Figure 6.1.6 depicts the process of transpiration. **Explain** how this process interacts with the hydrosphere, the geosphere, the biosphere and the atmosphere.

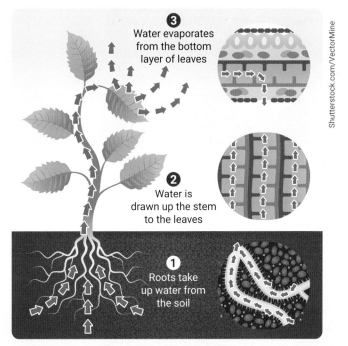

Shutterstock.com/VectorMine

❸ Water evaporates from the bottom layer of leaves

❷ Water is drawn up the stem to the leaves

❶ Roots take up water from the soil

▲ **FIGURE 6.1.6**   The process of transpiration

5 **Compare** the hydrosphere with the atmosphere, using their definitions.

6 Is a volcanic eruption a good example of different spheres interacting? **Justify** your answer by drawing a labelled diagram.

✓ state the word and chemical equations of photosynthesis and cellular respiration

✓ describe the location and purpose of photosynthesis and cellular respiration

✓ explain why photosynthesis can only occur during the day.

GET THINKING

You may sometimes feel very energetic. What is the source of your energy? Perhaps a sandwich or a glass of juice? And what is the original source of that food or drink? The answer is the Sun. Write two or three sentences explaining what you think the link is between the Sun's energy and the energy you get from food and drink. Then check your thinking at the end of the module.

producer
an organism such as a plant that converts light energy into chemical energy

organic compound
a compound containing carbon, usually bonded to other carbon and hydrogen atoms

chloroplast
a cell organelle in producers that absorbs the light energy for photosynthesis

chlorophyll
a chemical pigment, usually green, found in chloroplasts

Photosynthesis and cellular respiration are two important processes that circulate carbon around Earth's spheres.

Photosynthesis

Some of the energy that enters the Earth system is in the form of light energy from the Sun. Photosynthesis is a process in which **producers** use light energy, carbon dioxide and water to produce sugars (**organic compounds**) such as glucose. Plants, algae and some bacteria are producers. They transform light energy into a form of potential chemical energy, stored inside organic molecules such as glucose. Because photosynthesis needs sunlight, it can only occur during the day.

Photosynthesis takes place inside cell organelles called **chloroplasts**. These contain **chlorophyll**, a chemical pigment that absorbs light energy for producers to use. Chloroplasts occur in leaf cells of plants and within some bacteria cells. Chlorophyll is located in the membranes of thylakoids, which are compartments inside chloroplasts.

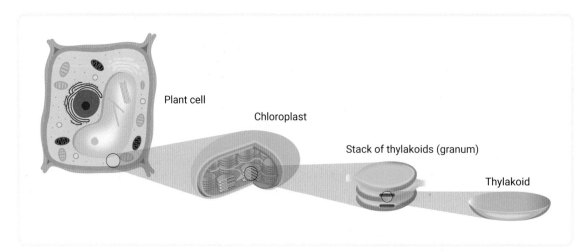

▲ **FIGURE 6.2.1** A stack of thylakoids inside the chloroplast of a cell, each containing chlorophyll

The process of photosynthesis can be summarised as a word or chemical equation, as shown in Figure 6.2.2.

a   Carbon dioxide + Water $\xrightarrow[\text{Chlorophyll}]{\text{Light}}$ Glucose + Oxygen

b $6CO_2$ + $6H_2O$ $\xrightarrow[\text{Chlorophyll}]{\text{Light}}$ $C_6H_{12}O_6$ + $6O_2$

▲ FIGURE 6.2.2   (a) A word equation and (b) a simple chemical equation for photosynthesis

Much of the sugar (glucose) that producers make during photosynthesis is stored. The chemical energy stored in glucose is then transformed into a useable form in the process of **cellular respiration**.

Cellular respiration and ATP

Cellular respiration occurs when an organism needs to use the stored energy in its cells to perform life processes such as growth, movement and reproduction. A producer does this by breaking the chemical bonds in the sugars it has made during photosynthesis. This releases energy in the form of **adenosine triphosphate (ATP)**. Consumers get their sugars by eating food, instead of by photosynthesis. But the process of cellular respiration is the same.

As an organism uses up energy, it converts ATP to adenosine diphosphate (ADP). As chemical bonds are broken and formed, phosphate (P) molecules and energy are released and absorbed, as shown in Figure 6.2.3. ATP (three phosphate molecules) has more stored energy than ADP (two phosphate molecules) because it stores energy in the extra phosphate–phosphate bond.

The ATP–ADP cycle occurs inside the cells of most living things.

Types of cellular respiration

There are two types of cellular respiration: aerobic respiration and anaerobic respiration.

- Aerobic cellular respiration occurs in the presence of oxygen in cell organelles called mitochondria (singular: **mitochondrion**), often described as the 'powerhouse' of a cell (Figure 6.2.4). This type of respiration produces more ATP than anaerobic respiration does.

cellular respiration
the process in which producers and consumers break down organic compounds (e.g. glucose) to produce ATP (energy)

adenosine triphosphate (ATP)
a useable form of energy in living things

mitochondrion
an organelle where aerobic cellular respiration occurs and produces a useable form of energy

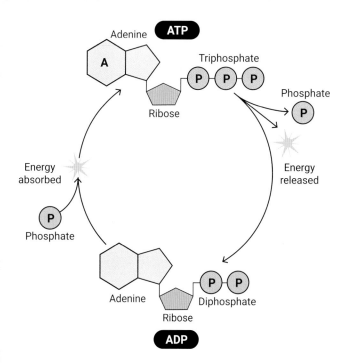

▲ FIGURE 6.2.3   The ATP–ADP cycle

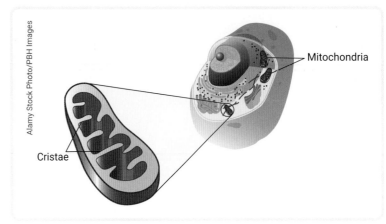

▲ FIGURE 6.2.4   A mitochondrion enlarged and its position in a cell's cytosol

cytosol
the liquid substance
inside a cell that holds
organelles in place

- Anaerobic respiration occurs when there is no oxygen. This type of respiration takes place in the **cytosol**, the liquid inside a cell contained by the cell membrane. Anaerobic respiration produces less ATP than aerobic respiration.

Once a glucose molecule has been broken down into smaller molecules, they can move from the cytosol into the mitochondrion if oxygen is present. Inside a mitochondrion, the inner membrane contains many folds, known as cristae, as shown in Figures 6.2.4 and 6.2.5. Cristae increase the surface area of mitochondria, which allow them to produce more ATP more quickly.

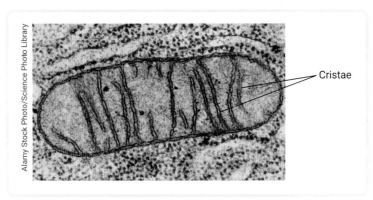

▲ FIGURE 6.2.5   A electron micrograph of a mitochondrion

The word equation and chemical equation for cellular respiration are shown in Figure 6.2.6.

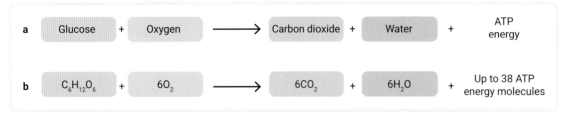

▲ FIGURE 6.2.6   (a) A word equation and (b) a simple chemical equation for cellular respiration

Cellular respiration in a bag

Aim

To model the process of cellular respiration

You need

- zip-lock bag
- packet of yeast
- ½ cup of warm water
- ½ cup of crushed sugary cereal

What to do

1 Put the cereal in the zip-lock bag.
2 Crush the cereal.
3 Add the yeast to the bag.
4 Add the water and then seal the bag shut.
5 Leave for about 60 minutes and observe what happens.

What do you think?

1 What happened to the bag?
2 What do you think is causing this?
3 How do you think this activity relates to the process of cellular respiration?

6.2 LEARNING CHECK

1 **State** the word equation for photosynthesis.
2 **State** the chemical equation for photosynthesis.
3 **Describe** the function of chlorophyll in photosynthesis.
4 **Explain** why photosynthesis and cellular respiration are both necessary in plants.
5 Look at Figure 6.2.7 and then draw a Venn diagram to **compare** photosynthesis and cellular respiration.

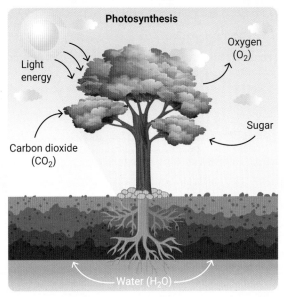
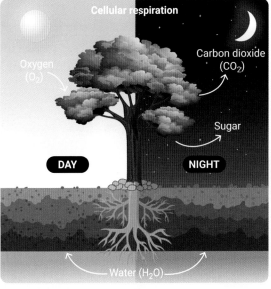

▲ **FIGURE 6.2.7** Comparing photosynthesis and cellular respiration

BY THE END OF THIS MODULE, YOU WILL BE ABLE TO:

✓ define combustion, fossil fuels, natural gas, coal, crude oil and hydrocarbon
✓ write a generalised word equation for the combustion of hydrocarbons
✓ state the steps involved in the formation of fossil fuels and explain why they are non-renewable.

GET THINKING

Prepare a concept map to summarise the content in Modules 6.1 and 6.2. As you work through Modules 6.3–6.5, add information and include diagrams to help your understanding.

fossil fuel
a substance containing hydrocarbons that is used as an energy source; takes millions of years to form from the remains of dead plants and animals

hydrocarbon
a substance containing the elements hydrogen and carbon

non-renewable
a substance that cannot be easily replaced at the same rate it is consumed

crude oil
an unrefined liquid form of a fossil fuel

natural gas
a gaseous form of a fossil fuel; usually methane, propane or butane

coal
a solid form of a fossil fuel

Combustion

Combustion is the third major process in the carbon cycle. Combustion is a fast chemical reaction that transforms carbon from the biosphere and geosphere into carbon dioxide, which is transferred back into the atmosphere.

Fossil fuels

Fossil fuels are substances containing **hydrocarbons** (which contain carbon and hydrogen) that are burned to produce energy. Fossil fuels form over millions of years from the remains of ancient living organisms. This process only occurs if:

- the organic matter, such as plants and algae, is rapidly buried after dying
- there is no oxygen, which stops the matter from decomposing
- the organic matter is covered with layers of sediment, usually from oceans, seas and swamps
- the matter is pushed far below the surface of Earth many millions of years later
- the pressure and heat at great depths transform the organic matter into fossil fuels.

Because they take so long to form, fossil fuels are **non-renewable**. This means they cannot be replenished as fast as they are used.

Fossil fuels can be classified into three main categories: **crude oil**, **natural gas** and **coal** (Figure 6.3.1). Crude oil is sometimes called petroleum.

▲ **FIGURE 6.3.1** **(a)** Crude oil, **(b)** natural gas and **(c)** coal are fossil fuels.

Crude oil and gas form from marine (ocean-dwelling) organisms, as shown in Figure 6.3.2. Coal is formed from swamp and forest organisms, as shown in Figure 6.3.3.

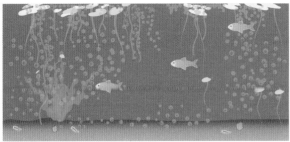

1 Tiny marine organisms, such as plankton and algae, die and sink to the bottom of the ocean.

2 The organisms are covered in mud, creating sediment. Oxygen levels remain low, preventing decomposition.

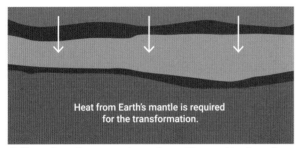

Heat from Earth's mantle is required for the transformation.

3 Over a long time, many layers form. The combination of heat from the mantle and pressure from the weight of the sediments above compress and transform the organisms into crude oil.

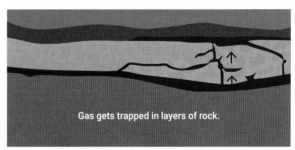

Gas gets trapped in layers of rock.

4 Oil moves up through cracks in sedimentary rock layers, forming a reservoir. Natural gas collects in hollow spaces. The temperature, pressure and type of organic material determine whether oil or gas is formed. Gas forms under higher temperatures and pressures.

▲ **FIGURE 6.3.2** The steps of crude oil and gas formation

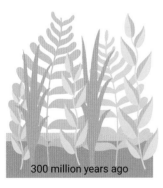

300 million years ago

1 Large numbers of large forest plants die in swamps.

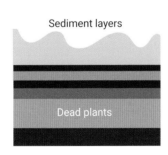

Sediment layers

Dead plants

2 Over millions of years, layers of sediment fall on top, increasing the weight and pressure on the buried plants.

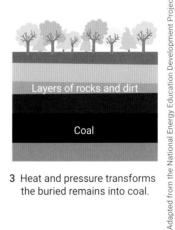

Layers of rocks and dirt

Coal

3 Heat and pressure transforms the buried remains into coal.

▲ **FIGURE 6.3.3** The steps of coal formation

Adapted from the National Energy Education Development Project

Video activity
Oxygen and combustion

Quiz
Fossil fuels

Australia has significant coal resources in the Bowen–Surat Basin, Queensland, and crude oil and gas resources in the Carnarvon Basin off the west and north-west coast of Western Australia. According to Geoscience Australia's 2019 assessment, Australia's natural gas deposits are expected to run out in approximately 34 years.

Crude oil

Crude oil is used to make petrol, diesel and jet fuel. Crude oil is brought out of the ground as a mixture of different hydrocarbons. Heavier components are used to make tar, asphalt, paraffin wax and lubricants. Other goods that can be made from crude oil include plastics, clothing and chewing gum.

Every day, humans use more than 70 million barrels of crude oil. One barrel contains enough hydrocarbons to make 540 plastic toothbrushes. That's the equivalent of almost 40 billion plastic toothbrushes of crude oil produced each day!

Combustion

Many human activities involve the combustion of fossil fuels, such as burning coal, natural gas and crude oil. Activities such as the burning of wood also produce carbon dioxide. In Australia, most of our power stations, manufacturing, construction industries and transport use combustion.

▲ **FIGURE 6.3.4**   Combustion of natural gas

In Chapter 5, we looked at combustion chemical reactions. You learned that combustion is an exothermic reaction because it releases energy. We also looked at the two types of combustion reactions: complete combustion and incomplete combustion. In this module, we will just look at complete combustion.

When a substance such as a fossil fuel is burned in oxygen, it releases energy, carbon dioxide and water.

When different fossil fuels are burned, the same products are released. That is because fossil fuels are mostly made of hydrocarbons, compounds that are composed only of hydrogen and carbon. A combustion reaction can be shown as a word equation:

fossil fuel + oxygen → carbon dioxide + water

The carbon in fossil fuels was taken out of the atmosphere millions of years ago by producers via photosynthesis. All living organisms gain this carbon. Producers gain carbon through photosynthesis, and consumers gain carbon by eating producers. When we burn fossil fuels, we are returning huge amounts of carbon to Earth's atmosphere in the form of carbon dioxide.

Distillation

Fractional distillation is a process used to separate crude oil into different, useful substances such as a petrol. In the school laboratory, we can perform a simplified version of fractional distillation, known as simple distillation. This is a separation technique. This can be a teacher demonstration or student activity.

Aim

To separate ethanol from an ethanol–water mixture by simple distillation

You need

- 100 mL of ethanol–water mixture (in a 1:4 ratio of ethanol to water, so 20 mL/80 mL)
- distillation apparatus (round-bottom flask, Bunsen burner or heating mantle, condenser and thermometer)
- thermometer
- Bunsen burner
- tripod stand and clamp to hold distillation flask
- small beaker to capture ethanol

What to do

1. Put 100 mL of ethanol–water mixture into the distillation (round-bottom) flask.

2. Set up the heating mantle or Bunsen burner and distillation apparatus, as shown in Figure 6.3.5.

3. Heat the distillation flask to maintain 78°C, until the ethanol has evaporated and then condensed and collected in the small beaker.

4. To confirm the substance you have collected, pour 5 drops of ethanol onto a watch glass to ignite it. Ethanol is flammable. Water is not flammable.

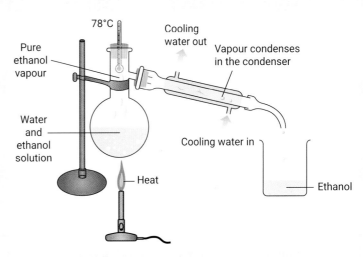

▲ **FIGURE 6.3.5** The experimental set-up

What do you think?

Describe the purpose of the condenser. Explain why the property, known as flammability, was used to confirm the separated substance.

6.3 LEARNING CHECK

1. **Define** 'combustion'.
2. **List** three substances that are made from crude oil.
3. **Determine** whether the following statements about the process of fossil fuel formation are true or false. Rewrite the false statements to make them true.
   a. The organic matter, such as plants and algae, is rapidly buried after dying.
   b. Oxygen must be present for the decomposition of plants and algae.
   c. The organic matter is covered with layers of sediment.
   d. The process requires hundreds of years.
   e. Friction, at great depths, transforms the organic matter into fossil fuels.

✓ define carbon, biogeochemical cycle, reservoir and carbon sink

✓ describe the carbon cycle and draw a labelled diagram to model the carbon cycle

✓ explain how the carbon cycle results from the interaction of Earth's four spheres.

Video activity
The carbon cycle

Interactive resource
Label: Carbon cycle

GET THINKING

Recall the three processes studied so far: photosynthesis, cellular respiration and combustion. They are the three main processes involved in the cycling of carbon around Earth's spheres. Think about how the reactants and products of the three processes interact. How does each process depend on the others?

Carbon compounds

Carbon is present in all organisms. Its unique properties allow it to bond easily with a variety of other atoms such as hydrogen and oxygen. Energy can be stored in these bonds. Large complex molecules such as the hydrocarbons in fossil fuels can store enormous amounts of energy.

Some forms of carbon found in the carbon cycle are:

- fossil fuels
- glucose, proteins, nucleic acids (e.g. DNA) and lipids found in living organisms (producers and/or consumers)
- carbon dioxide (see Figure 6.4.1)
- calcium carbonate
- carbon stored in rock as shale or coal
- wood.

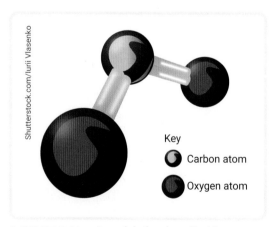

Key

◑ Carbon atom

● Oxygen atom

▲ **FIGURE 6.4.1** A model of carbon dioxide – a form of carbon in the carbon cycle

9780170472838

Biogeochemical cycles

A **biogeochemical cycle** is how scientists describe how chemical substances, such as carbon and water, are transformed, transferred and stored on Earth through natural processes.

Carbon is constantly moving between different forms through Earth's four spheres. It is stored in **reservoirs** for long periods of time. To move from one reservoir to another, carbon needs to go through processes such as photosynthesis, cellular respiration and combustion. Other processes that move carbon through the reservoirs are:

- **consumption**
- **decomposition** (Refer to Figure 6.4.2 to observe organic matter before and after composting.)
- fossilisation
- weathering and erosion
- excretion
- dissolving in oceans
- volcanic eruptions.

The four main reservoirs of carbon are the:

- atmosphere – mostly in the form of carbon dioxide gas
- biosphere – includes producers, consumers and non-living organic material such as decaying leaves
- oceans – includes dissolved carbon dioxide in the hydrosphere
- sediments – includes sedimentary rocks, fossil fuels and soil in the geosphere.

biogeochemical cycle
a model describing how chemical substances are transformed and stored in Earth's spheres

reservoir
a long-term storage area of carbon, such as the atmosphere, the biosphere, oceans and sediments

consumption
the process of a consumer feeding on a producer or another consumer, resulting in a transfer of energy and carbon

decomposition
breaking down complex organic matter into simple inorganic molecules in a form that can be taken up again by plants

▲ **FIGURE 6.4.2**   Organic matter before and after decomposition

The carbon cycle

The carbon cycle is the biogeochemical cycle that transfers and transforms carbon through Earth's biosphere, hydrosphere, geosphere and atmosphere (Figure 6.4.3). Carbon atoms circulate between biotic and abiotic components of Earth through a number of complex pathways, and together these form the carbon cycle.

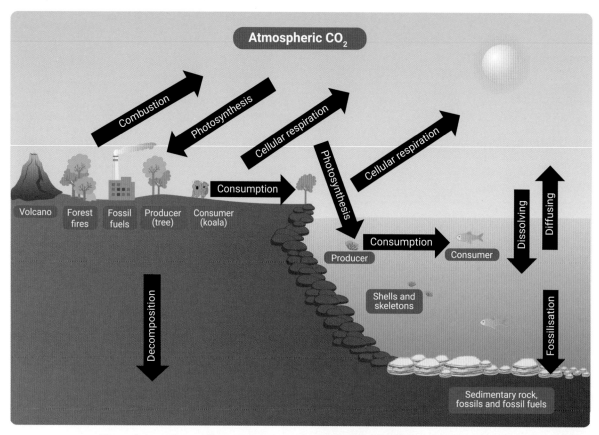

▲ **FIGURE 6.4.3** The carbon cycle – carbon circulates around Earth's four spheres.

carbon sink
a type of reservoir where more carbon is absorbed than is released

When carbon reservoirs are storing more carbon than they are releasing, they are known as a **carbon sink**. Common carbon sinks are rocks that contain carbon in the form of coal, forests that contain carbon in the form of stored sugars in plants (Figure 6.4.4), and the atmosphere that contains carbon in the form of carbon dioxide.

▲ **FIGURE 6.4.4** Tropical rainforests are a carbon sink.

Carbon cycles quickly and slowly

Carbon can cycle fast through the biosphere, within the life span of a human. As you saw in Module 6.2, producers transform carbon dioxide in the atmosphere into

9780170472838

glucose via the process of photosynthesis. Glucose is stored in producers. The carbon in glucose is transferred to consumers through consumption. All organisms transform the stored energy into a useable form during cellular respiration. During respiration, the carbon is released as carbon dioxide into the atmosphere. Bacteria contribute to this speedy cycle by consuming decaying organisms and releasing carbon dioxide as they respire (Figure 6.4.5).

Carbon cycles slowly through the geosphere. Carbon compounds in organisms can be transformed into fossil fuels over millions of years. People extract fossil fuels from the ground to produce energy for combustion engines and other activities. This returns carbon to the atmosphere in the form of carbon dioxide. When a volcano erupts, carbon is again released as carbon dioxide as it transfers from the geosphere to the atmosphere.

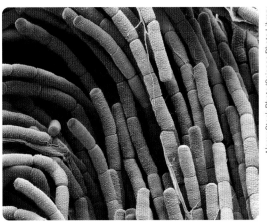

▲ **FIGURE 6.4.5** Bacteria from compost. In the process of composting, micro-organisms like these bacteria break down organic matter and produce carbon dioxide and water.

6.4 LEARNING CHECK

1 **List** three main processes involved in the carbon cycle.

2 **Describe** the fast carbon cycle.

3 **Define** 'carbon sink'.

4 **Draw** a diagram to represent the slow carbon cycle.

5 Human activities such as deforestation and burning fossil fuels put excess carbon dioxide into the atmosphere. **Survey** five students in your class to find out what activities they are involved in that puts more carbon dioxide into the atmosphere.

6 Figure 6.4.6 is a simplified representation of how carbon dioxide moves through the carbon cycle. **Create** your own version of the carbon cycle by copying the figure into your book and:

- adding labels to show what processes the arrows represent
- labelling Earth's four interacting spheres
- providing further information about how carbon cycles through the spheres.

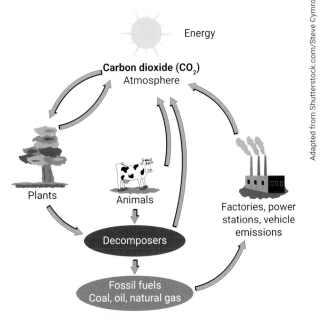

▲ **FIGURE 6.4.6**   The carbon cycle

BY THE END OF THIS MODULE, YOU WILL BE ABLE TO:

✓ define greenhouse gas, greenhouse effect and enhanced greenhouse effect

✓ relate changes to the carbon cycle to the greenhouse effect and explain how it leads to the enhanced greenhouse effect.

GET THINKING

Return to the concept map you started at the beginning of Module 6.3. Do you now have more information about the carbon cycle to add to the map? Can you draw a more complex map with more detail? Some terms to consider are 'decomposition', 'fossilisation' and 'consumption'.

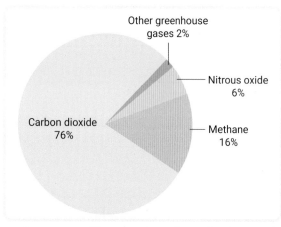

▲ **FIGURE 6.5.1** The proportions of greenhouse gases in the atmosphere

Greenhouse gases

Greenhouse gases are gases in the atmosphere that can absorb and trap heat energy. Water vapour, carbon dioxide, nitrous oxide and methane are examples of greenhouse gases (Figure 6.5.1). These are emitted naturally; for example, through volcanic eruptions and bushfires. Greenhouse gases such as carbon dioxide and methane are also produced and emitted by human activities, especially the burning of fossil fuels. The **greenhouse effect** is the heat-trapping effect of these gases in the atmosphere.

greenhouse gas
a gas in the atmosphere that can trap heat and affect global surface temperatures

greenhouse effect
a natural process that traps energy within the atmosphere, causing Earth's surface temperature to increase

The greenhouse effect

The process known as the greenhouse effect starts with the Sun's radiation, which includes visible light. Most visible light passes through Earth's atmosphere and hits Earth's surface. Some of it is absorbed and is re-emitted as longwave radiation (infrared radiation), which we feel as heat. You will learn more about visible light and longwave radiation in Chapter 7.

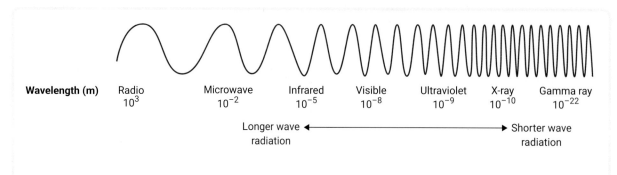

▲ **FIGURE 6.5.2** The wavelengths of different types of radiation

The radiation travels back through Earth's atmosphere, where it is absorbed and re-emitted by greenhouse gases. Reflection of the radiation from the gases back to Earth traps heat in the atmosphere (see Figure 6.5.3). This trapping effect insulates Earth, maintaining average global temperatures that are hospitable to life. Without the greenhouse effect, Earth's average global temperatures would be below freezing.

The **enhanced greenhouse effect** is the term used to describe the increased warming effect we are experiencing because of the rapid increase in greenhouse gases. This enhanced greenhouse effect is directly caused by an increase in greenhouse gas levels due to human activities.

As we emit more greenhouse gases such as carbon dioxide into the atmosphere, the enhanced greenhouse effect is increasing. This in turn is causing average temperatures on Earth to increase.

enhanced greenhouse effect
the increased warming of the atmosphere and Earth's surface due to greenhouse gases produced by human activities

Video activities
The greenhouse effect
Climate change

Interactive resources
Simulation: Greenhouse effect
Drag and drop: Natural v enhanced greenhouse effect

Extra science investigation
Greenhouse effect in a bottle

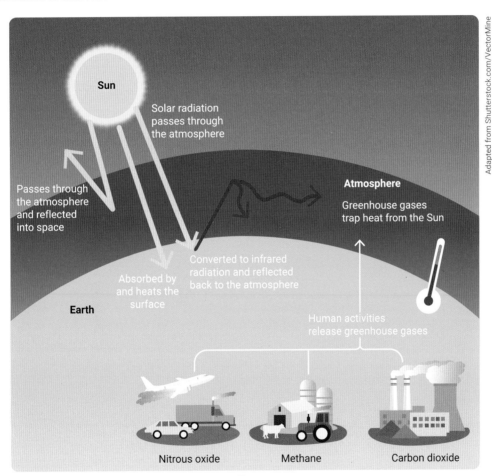

▲ **FIGURE 6.5.3** The enhanced greenhouse effect

Changes to the carbon cycle

The carbon cycle demonstrates pathways of carbon in and out of the atmosphere. The rate at which carbon is being emitted into the atmosphere has increased significantly since the Industrial Revolution in the mid-1700s. More agriculture and industry and an increasing global population means more people need feeding and more energy is needed for transport and electricity. As a result, we are directly causing changes to the carbon cycle by activities such as burning fossil fuels for energy. At the same time, we are depleting the carbon reservoirs such as forests through activities such as land clearing for agriculture and resources. This means less carbon dioxide is being absorbed by carbon reservoirs.

Evidence of change in the carbon cycle

Scientists have data showing thousands of years of natural atmospheric carbon dioxide fluctuations of highs of about 280 parts per million (ppm) and lows of about 200 ppm. However, more recent data shows increases in carbon dioxide that have never been recorded before. For example, in 2013 the maximum concentration was a new record of 400 ppm. As of 2022, the concentration of carbon dioxide is predicted to pass 420 ppm. Figure 6.5.4 illustrates this sudden rise.

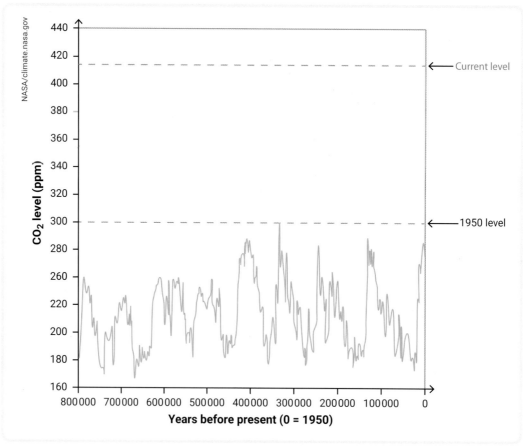

▲ **FIGURE 6.5.4** Changes in global atmospheric concentrations of carbon dioxide. For thousands of years, carbon dioxide levels did not rise above 300 ppm.

Photo by QUI NGUYEN on Unsplash

▲ **FIGURE 6.5.5** Corals need calcium carbonate for their skeletons. Increased levels of carbon dioxide in oceans makes the water more acidic. This makes it harder for corals to build their skeletons.

Scientists use high-performance **supercomputers** to predict future changes to the carbon cycle, such as increases in atmospheric carbon dioxide. Supercomputers are extremely powerful computers. They use multiple processors that run in parallel and enable a very fast processing speed. Australia's Commonwealth Scientific and Industrial Research Organisation (CSIRO) has access to computers with the capability of one quadrillion (million billion) operations per second.

supercomputer
a powerful computer with multiple processors that run in parallel, significantly faster than a single computer

The World Heritage-listed Great Barrier Reef is threatened by warming oceans and increasing carbon dioxide levels. Scientists use sensors to monitor changes in oceans that will affect their ability to store carbon. For example, a sensor is attached to the cargo ship RTM *Wakmatha*. As the ship travels along the reef from Weipa to Gladstone, the sensor records carbon dioxide levels. More than 600 species of coral are susceptible to changes in pH, a measure of acidity in the water. As more carbon dioxide dissolves in the ocean, the pH decreases, indicating a higher acidity. This can prevent a build-up of calcium carbonate. Corals require calcium carbonate, usually available in the ocean, to build their skeletons (Figure 6.5.5).

6.5 LEARNING CHECK

1 **List** four greenhouse gases.
2 **Identify** the different types of radiation involved in the greenhouse effect.
3 **Predict** the impact on the carbon cycle of too little carbon dioxide in the atmosphere.
4 **Explain** why the greenhouse effect benefits humans.
5 **Research** different ways you can model the greenhouse effect in the classroom with containers. Or try the PhET simulation at the weblink.

Weblink
PhET: Greenhouse effect

✓ describe some of the ways carbon dioxide can be captured and stored.

Video activity
What is carbon
farming?

**Extra science
investigation**
Effect of CO_2 on water

GET THINKING

Every day, you make many choices that affect how much carbon dioxide and other greenhouse gases are produced. Write a list of some of these choices and then share your list with a partner. Which of these choices do you think could have the most positive impact on the environment? Discuss your reasoning with your partner.

▲ **FIGURE 6.6.1** Planting more trees is one way to reduce the concentration of carbon dioxide in the atmosphere.

carbon sequestration
the process of capturing and storing atmospheric carbon dioxide

Carbon sequestration

With human activities emitting more carbon dioxide than can be absorbed by Earth, the concentrations of carbon dioxide in the atmosphere are increasing. In Module 6.4, we looked at the role of Earth's carbon reservoirs, which absorb and store vast amounts of carbon. More than ever, these reservoirs are key to stabilising Earth's carbon dioxide levels and reducing the enhanced greenhouse effect.

Carbon sequestration is a method that reduces the concentration of carbon dioxide in the atmosphere. It can occur naturally or artificially. Natural methods to increase the rate of carbon sequestration include reforestation, (Figure 6.1.1) revegetation and restoring land used for farming to increase the rate of photosynthesis.

An artificial method involves capturing carbon dioxide produced by coal-powered plants before it enters the atmosphere. The captured gas can then be stored in soil or oceans or injected into underground rock, as shown in Figure 6.6.2.

Carbon farming is another method of reducing atmospheric carbon. It is focused on improving plant and soil management. Usually, the purpose

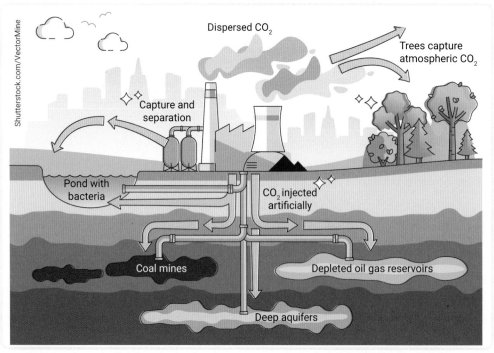

Dispersed CO_2

Trees capture atmospheric CO_2

Capture and separation

Pond with bacteria

CO_2 injected artificially

Coal mines

Depleted oil gas reservoirs

Deep aquifers

▲ **FIGURE 6.6.2** Artificial methods of carbon sequestration include injecting carbon dioxide underground.

of farming is to harvest and use crops. In contrast, the purpose of carbon farming is to capture and store carbon. Farmers plant deeper-rooted plants and incorporate organic materials into the soil.

Other methods to reduce carbon dioxide

To reduce the carbon dioxide in the atmosphere, we all need to find ways to reduce the emissions produced by our activities, such as electricity generation and use of fuel. One way to do this is to use and invest in alternative, renewable energy sources such as solar and wind, which reduces emissions significantly. Another alternative is **clean hydrogen**, which the Australian Government is investing in as a future energy source (Figure 6.6.3). When clean hydrogen is burned, the only product is water.

There are also a lot of small steps you can take to reduce your own **carbon footprint**, starting today, such as carpooling, using public transport or riding your bike, using recycled paper, buying second-hand products and reducing your electricity consumption.

clean hydrogen
hydrogen that is produced with low or no carbon emissions

carbon footprint
the amount of carbon dioxide emitted by a person, organisation or event

6.6

Shutterstock.com /Scharfsinn

▲ **FIGURE 6.6.3**   Hydrogen fuel cells can be used to power buses.

Weblink
Department of Industry,
Science, Energy and Resources

6.6 LEARNING CHECK

1 Look at Figure 6.6.4 to work out which of your own activities produce carbon dioxide emissions. **Write** down five activities and then suggest how you could change them to reduce your carbon footprint.

2 Watch the video from Australia's Commonwealth Department of Industry, Science, Energy and Resources at the weblink to learn about the future of clean hydrogen use in Australia.

   a **List** the positives of using clean hydrogen as an energy source.

   b **Write** four or five questions or points about hydrogen energy you could investigate further. Then use this information to answer the question 'Are there any negative aspects to using hydrogen as a fuel?'

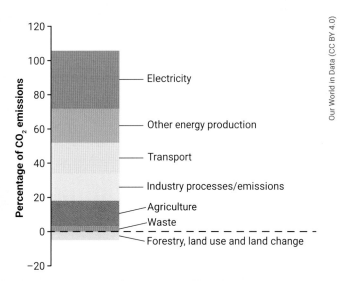

▲ **FIGURE 6.6.4**   Australia's carbon dioxide emissions by source, 2020 (based on figures from the National Greenhouse Gas Inventory)

6.7 First Nations Australians' fire management of Country

IN THIS MODULE, YOU WILL:

✓ investigate the impact of environmental fire management of Country on reducing carbon emissions.

Managing the environment with fire

Uncontrolled fires in Australia in the last 200 years have cost more than $1.6 billion. Uncontrolled fires have devastating impacts on properties, crops and livestock. They endanger Australia's biodiversity and increase carbon emissions.

First Nations Australians have used fire to manage the environment for thousands of years before European colonisation.

Early Europeans misunderstood fire management by First Nations Australians, thinking it was dangerous and destructive, and they prevented the traditional-style burning of the environment. This change in management altered the environment. Introduced weeds and native grasses have been left to grow uncontrollably, while soil erosion and soil salinity have increased. These changes have culminated in an increase in uncontrolled wildfires.

Now, contemporary science is working with traditional management practices of First Nations Australians to manage the environment. Carefully placed and controlled burns can positively affect the environment, including reducing the risks of wildfires and lowering greenhouse gas emissions (Figure 6.7.1).

▲ **FIGURE 6.7.1** (a) A low-intensity burn on Lardil Country (Mornington Island, North Queensland); (b) grass and trees after a low-intensity burn

A case study

The North Australian Indigenous Land and Sea Management Alliance has been working to demonstrate the impact of traditional-style burning. Teams of First Nations land managers and science partners select regions of the land to study. They record the types of plants on the ground and in the canopy, as well as the grass cover and the height and width of trees. They measure a number of 1 m² plots, collecting and weighing the grass and ground litter. This is the 'before burning' measurement.

Carefully controlled fires are then set to remove the grasses and other fuels on the ground. The sampling is repeated after burning to measure the amount of fuel that has burned. Measurements of the smoke produced provide information about the gases released by burning. The same process is carried out in areas affected by wildfire.

☆ ACTIVITY

Investigating the impact of fire management

Projects such as the ones mentioned in this module demonstrate the potential to reduce harmful greenhouse gases.

1 Table 6.7.1 shows calculations for Matuwa (Lorna Glen National Park in Western Australia). This region, the lands of the Martu Peoples in the desert area, is dominated by dry spinifex grass. The Martu Peoples have Native Title rights to about 20 million hectares of north-western Western Australia. The calculations assume about 7 per cent of the land is managed by fire each year. Martu Peoples set fires in small patches to protect the habitats of native animals. This is known as mosaic burning because of the patch-like pattern seen on the land.

▼ TABLE 6.7.1 Calculations of fuel and emissions for traditional-style fire management versus those for unmanaged fires in Matuwa

	Fuel burned (t ha^{-1} year)	CH$_4$ emissions (t ha^{-1} year)	NO$_2$ emissions (t ha^{-1} year)	CO$_2$ emissions (t ha^{-1} year)
First Nations (Martu Peoples) traditional-style fire management	5.02	0.0095	0.0003	0.281
Unmanaged fires	6.64	0.0126	0.0003	0.371

Data from https://nailsma.org.au/uploads/resources/Desert-Fire-and-Carbon-Report-230714.pdf

a Calculate the difference in potential emissions of each of the gases and the amount of fuel burned.

b What is the difference in emissions between fire-managed land and unmanaged land?

c Martu Country comprises 20 million hectares of land. How does this affect the gas emissions according to the information in Table 6.7.1?

2 Figure 6.7.2 shows areas of Martu Country affected by fire in 1954, 1973 and 2000. Martu Peoples had been absent from their Country from the 1960s and through Native Title determination have returned since 2000.

1954: Martu present   1973: Martu absent   2000: Martu present

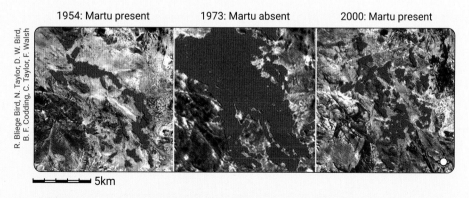

R. Bliege Bird, N. Taylor, D. W. Bird, B. F. Codding, C. Taylor, F. Walsh

├────┼────┤ 5km

▲ FIGURE 6.7.2   Areas affected by fire in Martu Country. The middle image was taken when no Martu Peoples were on Country.

Explain why you think there are differences in the area affected by fire on Martu Country. Use information from Table 6.7.1 to help you construct an explanation.

6.8 Why CSIRO monitors greenhouse gases at Kennaook/Cape Grim

BY THE END OF THIS MODULE, YOU WILL BE ABLE TO:

✓ interpret and analyse data to make decisions based on the needs of society

✓ investigate how the values and needs of society influence the focus of scientific research

✓ investigate how the need to minimise greenhouse gas emissions has led to scientific and technological advances.

Video activity
Cape Grim monitoring station

Kennaook/Cape Grim air monitoring site

Greenhouse gases are on the rise because of human activities. Scientists have shown that there is a clear link to this fast rise in the atmospheric concentration and the enhanced greenhouse effect. This is leading to global warming and other changes to the climate.

In the early 1970s, Australia's CSIRO made a commitment to the United Nations Environment Programme to monitor greenhouse gases in the atmosphere for the purposes of climate change research. The monitoring location is Kennaook/Cape Grim, a site on the north-west tip of Tasmania (Figure 6.8.1). This location was chosen because Kennaook/Cape Grim has some of the cleanest air in the world because the air comes from thousands of kilometres of ocean to the west and south-west. Data from Kennaook/Cape Grim Baseline Air Pollution Station is considered 'baseline', which allows scientists to compare current data to historic data sets and get a clear picture of global changes to the atmosphere. CSIRO has been continuously measuring the air at Kennaook/Cape Grim since 1976.

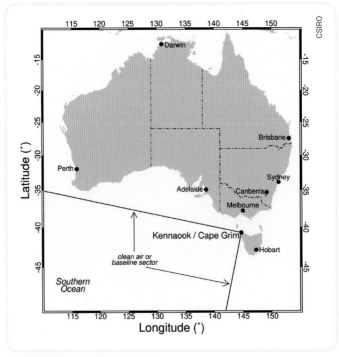

▲ **FIGURE 6.8.1** Kennaook/Cape Grim, on the north-west tip of Tasmania, is the site of CSIRO's monitoring station for air pollution.

What is being measured?

CSIRO uses very precise instruments to monitor more than 80 greenhouse gases at Kennaook/Cape Grim, including carbon dioxide and methane (CH_4). Scientists analyse and use the data to improve climate models, which helps them make more accurate predictions about climate change and weather patterns.

Since 1978, the atmospheric methane concentration has increased by about 24 per cent. The increase is attributed to human activities such as coal mining, use of landfills and agricultural practices.

Before the Industrial Revolution (mid-1700s), the concentration of methane in the atmosphere was 700 parts per billion (ppb). This means a dry sample of air containing 1 billion molecules would include 700 methane molecules. Today the value exceeds 1800 ppb.

Figure 6.8.2 shows atmospheric methane concentrations (in ppb) over the last 2000 years. This is based on measurements of air trapped in Antarctic ice and data from Kennaook/ Cape Grim.

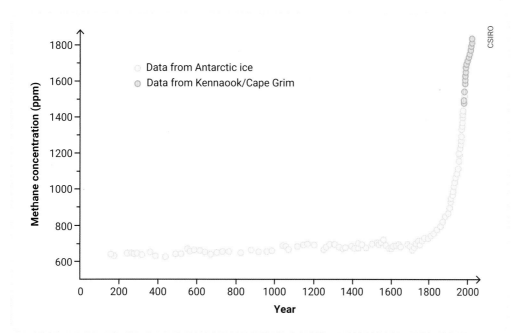

▲ **FIGURE 6.8.2** Atmospheric methane concentrations (in ppb) over the last 2000 years

CSIRO's Kennaook/Gape Grim data also shows that atmospheric concentrations of carbon dioxide have increased by over 25 per cent since 1976 (Figure 6.8.3).

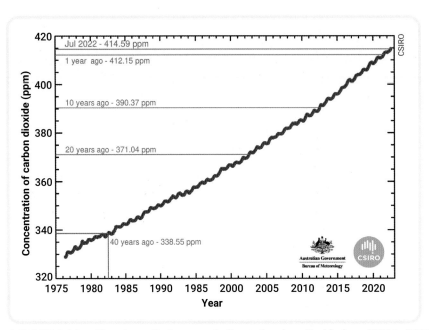

▲ **FIGURE 6.8.3**　The atmospheric concentrations of carbon dioxide from 1976 to 2022

How do scientists use the data?

The data from Kennaook/Cape Grim is available on CSIRO's website and from major global data archives. CSIRO's data is used as part of international assessments of climate change, such as those conducted by the Intergovernmental Panel on Climate Change (IPCC) and in Australia's *State of the Climate* reports.

The Australian Government also uses data from Kennaook/Cape Grim to meet international obligations and report on pollution levels. Scientists also use Kennaook/Cape Grim data in hundreds of research papers on climate change and atmospheric pollution.

6.8 LEARNING CHECK

1 **State** three reasons why CSIRO monitors greenhouse gases.
2 **Explain** why Kennaook/Cape Grim was chosen as an important site to conduct air monitoring.
3 Look at Figure 6.8.2. **Describe** the trend in atmospheric methane concentrations and **explain** how this relates to human activities.
4 Look at Figure 6.8.3. What do you notice about how atmospheric concentrations of carbon dioxide are increasing? Does it look as though carbon dioxide levels are increasing at a constant rate?

9780170472838

Reproducible methods and valid results

SCIENCE SKILLS IN FOCUS

IN THIS MODULE, YOU WILL FOCUS ON LEARNING AND IMPROVING THESE SKILLS:

▶ conducting a valid, reproducible investigation

▶ formulating and testing a hypothesis

▶ developing risk assessments and recognising potential hazards

▶ using multimodal/digital platforms to model and communicate the carbon cycle and display the processes.

▶ Reproducible methods and valid results

A scientific investigation needs a good, reproducible method to produce valid and reliable results. But what do 'valid' and 'reproducible' actually mean? And what should you do to ensure your methods are up to scratch?

A reproducible method is one where all the steps of an investigation, such as preparing, mixing, observing and measuring, can be repeated to produce the same or very similar results in the same circumstances. In other words, someone else could replicate your methods to get comparable and reliable results.

Valid means that the actions (method), data (results) and inferences (conclusions) you produce are accurate and measure or show what you intended them to measure or show.

To get valid and reliable results, make sure your method is reproducible and valid by considering the following.

▶ Validity

1 Have a clear aim. What are you testing for? What do you want to find out?

2 Identify which variables you will control (dependent variable) and which variable you will change (independent variable).

Video
Science skills in a minute: Reproducible methods and accurate results

Science skills resource
Science skills in practice: Reproducible methods and accurate results

3 Use a control group – a group in which the independent variable is not included. This is what you compare your other results with.

▶ Accuracy

4 Write the method clearly, step by step.

5 Be precise about what you will measure or observe, and how you will do it.

6 Always state the units of measurement (e.g. seconds, millilitres, grams).

▶ Reliability

7 Use large sample sizes or multiple trials/tests to get as much data as possible.

8 Calculate averages from repeated trials or tests to get a more accurate result (e.g. the sum of the mass of 30 apples divided by the total number of apples gives you the average mass).

INVESTIGATION 1: MODELLING OCEAN ACIDIFICATION

AIM

To conduct an investigation to show the effect of carbon dioxide on the pH of seawater

BACKGROUND INFORMATION

Carbon dioxide reacts with seawater to produce carbonic acid.

You can determine the pH of a substance by checking the coloured chart that comes with universal indicator.

MATERIALS

☑ small-to-medium conical flask

☑ 50 mL measuring cylinder

☑ 40 mL of dilute NaOH solution

☑ 40 mL of universal indicator

☑ reuseable straw (metal straws can be put into dishwasher) or paper straws

☑ safety glasses

Warning

Be careful – acids can splash and burn eyes and skin.

Only blow through straw – do not suck because this may draw acid into your mouth.

METHOD

1 Determine your hypothesis by using an 'If …, then …' statement; for example, 'If I blow carbon dioxide into the solution continuously for 1 minute, then the colour will change to yellow, indicating a pH of 6'.

2 Put on safety glasses.

3 Measure 40 mL of NaOH solution with a measuring cylinder and transfer it to a conical flask.

4 Rinse the measuring cylinder with water and then add 3–4 drops of universal indicator. Add this to the conical flask.

5 Create a table, similar to Table 6.9.1, to record your observations.

▲ **FIGURE 6.9.1** Universal indicator turns pink in a solution of NaOH.

6 Submerge part of the straw and gently blow air into the solution in the conical flask for 1 minute. Repeat the experiment two or three times to find out if the result is reproduced consistently. Remember to keep all other variables the same.

7 Replicate the experiment by asking other students to do the same experiment.

RESULTS

Copy and complete Table 6.9.1 in your workbook.

▼ **TABLE 6.9.1** Results table

Experiment	Colour before blowing in the straw	Colour change(s) after 30 s of blowing in the straw	Colour after 1 min
1			
2			
3			
4			

EVALUATION

1 **Identify** the part of your experiment that represented seawater.

2 **Identify** the component of your experiment that represented carbon dioxide emissions.

3 After you complete your table of results, write a paragraph to summarise your findings about the colour of universal indicator under acidic and basic conditions.

4 Did you have to alter your hypothesis or method? Why?

CONCLUSION

Write a conclusion for the investigation. Include a statement about whether or not the results supported your hypothesis.

INVESTIGATION 2: MODELLING AND COMMUNICATING THE CARBON CYCLE

AIM

To create an original model of the carbon cycle using multimodal or multimedia platforms

METHOD

1. Decide who your audience is. Who are you creating this model for? It's important to tailor your model to the age, prior knowledge and interests of your audience.

2. Determine the purpose of your model. What specific things do you want to communicate to your audience? What do you want them to understand or learn about the carbon cycle?

3. Decide how you want to engage your audience's interest. Try using different ways (visual, audio, hands-on etc.) to grab your audience's attention.

4. Choose a program or platform to create your model, such as PowerPoint, Canvas, Google Slides or Prezi. You could also try to make it interactive.

5. Present your model to your target audience.

EVALUATION

1. Ask your audience for feedback about your model and presentation. For example, you could ask:
   - What did they like about the model and the presentation?
   - What didn't they like?
   - How could you improve the clarity or accuracy of the model?
   - How could you improve the mode or style of your presentation?
   - What did they learn about the carbon cycle from watching or interacting with your model?

2. Compare your feedback to the purpose you outlined in step 2 of the method. Did your model achieve its purpose?

INVESTIGATION 3: OBSERVING CELLULAR RESPIRATION WITH SEEDS

AIM

To observe cellular respiration and visualise the waste product CO_2 by using bromothymol blue indicator

BACKGROUND INFORMATION

Carbon dioxide (CO_2) makes solutions more acidic (lower pH). Bromothymol blue is an indicator that turns yellow in acids (including solutions containing CO_2) and blue in basic solutions (higher pH) (including solutions that don't contain CO_2).

Warning
- Wear appropriate personal protective equipment (PPE).
- Know and follow all regulatory guidelines for the disposal of laboratory wastes.
- Avoid any direct contact with the bromothymol blue and sodium hydroxide solutions.
- Wash your hands thoroughly before and after handling any chemicals.
- Sterilise work surfaces before and after the investigation.
- Do not eat or drink any of the materials in this investigation.

MATERIALS

☑ hydrated seeds (e.g. wheat, barley, peas or mung beans)
☑ sealable bag
☑ bromothymol blue solution (0.04%)
☑ sodium hydroxide solution (1%)
☑ distilled water
☑ 4 test tubes
☑ 2 test-tube racks
☑ rubber stoppers
☑ conical flask (250 mL)
☑ paper towel

1 Make a prediction. What do you think will happen to the seeds kept in the dark compared with the seeds kept in a bright location? Record your prediction in your results table (Table 6.9.2).

2 Prepare the bromothymol blue solution by adding 3 mL of 0.04% bromothymol blue to 160 mL of distilled water. The solution should turn green as the water is added. However, the colour of your solution may vary depending on the pH of your water.

3 This experiment works best when the solution begins slightly basic. Add 30–40 drops of sodium hydroxide solution until the colour changes from green to deep blue.

4 Three-quarters fill two test tubes with hydrated seeds.

5 Fill the rest of the space in the test tubes with bromothymol blue solution.

6 Seal the tubes with rubber stoppers.

7 Completely fill the final two test tubes with bromothymol blue solution and seal them with rubber stoppers. These will be the control test tubes.

8 Place one of the control test tubes and one of the test tubes with seeds in a sunny spot in the classroom.

9 Place the remaining two test tubes in a dark place, such as a cupboard, to control for the effects of photosynthesis.

10 Leave the test tubes overnight. Consider what you think the results might be and write your prediction in Table 6.9.2.

11 The next day, observe any colour changes in the solution and record your findings in Table 6.9.2.

RESULTS

Copy and complete Table 6.9.2 in your workbook.

▼ TABLE 6.9.2 Results table

Condition	Colour	
	Day 0	Next day
Light – seeds		
Light – control		
Dark – seeds		
Dark – control		

Prediction:

EVALUATION

1 Did your samples change colour? What does this indicate about the waste products of cellular respiration?

2 Was your prediction correct?

3 What effect did putting the sample in the dark have on the colour?

4 What does this observation indicate about a key difference between cellular respiration and photosynthesis?

5 What effect would removing seeds have on the carbon cycle?

6 Why did you use control samples?

1 **Define**:
   a biosphere.
   b geosphere.

2 **Name** three components of the geosphere.

3 **Write** the word equation for cellular respiration.

4 **Write** the simple chemical equation for cellular respiration.

5 **Define** 'fossil fuel'.

6 **Describe** the chemical relationship between photosynthesis and cellular respiration in terms of their reactants and products.

7 **Identify** the following statement as true or false. Coal is normally a solid fossil fuel, originally made of ancient plants that died in swamps.

8 **Identify** five processes that are part of the carbon cycle.

9 **Describe** the role of a volcano in the carbon cycle.

10 **Summarise** the benefits of using renewable energy sources instead of fossil fuels.

11 **Describe** three actions you could take to contribute to carbon sequestration.

12 **Describe** the effect of a natural event or process, such as an earthquake or plant decay, on the carbon cycle and how it could cause matter to move from one of Earth's spheres to another.

13 **Explain** why photosynthesis only occurs during the day.

14 **Explain** why crude oil is described as a non-renewable energy source.

15 **Explain** why carbon sequestration can help reduce the impact that humans have had on the carbon cycle.

16 **Explain** how the carbon cycle involves interactions between Earth's spheres.

17 **Create** a flow diagram that shows the connection between the Sun's energy and the greenhouse effect.

18 **Analyse** the following graph of atmospheric concentrations of carbon dioxide measured by CSIRO at Kennaook/Cape Grim Baseline Air Pollution Station see graph below). **Describe** the trend in concentrations over time and **link** this to human activities.

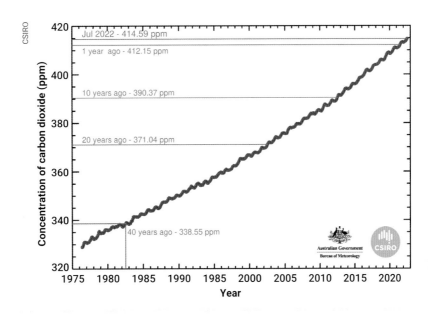

19 Analyse the following information about an extreme weather event and determine which of Earth's four spheres were interacting. **Justify** your answers.

In February 2022, very heavy rain fell over parts of eastern New South Wales and Queensland. Some areas received close to their annual total rainfall in just a few days. Heavy rain caused the breakdown of rock surfaces and landslides. Severe flooding occurred in many areas, affecting many thousands of people. Damage to houses, roads and the natural environment occurred. In Brisbane, more than a year's worth of waste resulted from flood damage to houses and businesses.

20 Rewrite the incorrect statement below to make it correct. **Justify** your changes.

First Nations Australians' traditional fire management practices are better for the environment because regular, high-intensity fires encourage the growth of weeds and grasses, and produce more methane and carbon dioxide.

21 Complete the summary table below to **connect** some of the pathways and processes in the carbon cycle.

22 Analyse the following model of the cycle of materials through the interacting organelles mitochondrion and chloroplast. **Relate** their processes to the carbon cycle by **drawing** and labelling your own diagram, or by **creating** a digital model using a program such as Prezi.

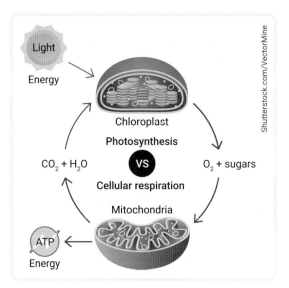

Sphere – carbon reservoir	Form of carbon before process	Process	Form of carbon after process
Hydrosphere – ocean	Carbon dioxide	Dissolving from atmosphere into the ocean	Carbon dioxide
		Photosynthesis by marine producers; e.g. phytoplankton	Glucose
Biosphere – forests	Glucose	Consumption and digestion	Glucose, protein, lipids, nucleic acids
Atmosphere			
Geosphere – sediment			

BIG SCIENCE CHALLENGE PROJECT #6

1 Connect what you've learned

In this chapter, you have learned about the processes and pathways of the carbon cycle and how carbon is stored, cycled and transformed through Earth's four spheres. You have also learned about how human activities affect the carbon cycle.

Review the mind map you have completed to ensure it connects all the key information you have learned in this chapter. Where possible, refine your mind map by adding more connections and details, such as labelled diagrams and examples.

Save the planet - use eco transport

Shutterstock.com/Pavel Vinnik

2 Check your thinking

Explain to a friend why changes to the carbon cycle have led to a higher concentration of atmospheric CO_2 and the enhanced greenhouse effect. Based on what you have learned in this chapter, prepare a response to the incorrect argument that: 'Climate change is due to an increase in the Sun's energy'. To help prepare your argument, do some research about why this statement is incorrect.

3 Make an action plan

Make a list of strategies for reducing your carbon emissions. Some of these strategies could be achieved in a few days, whereas others may take longer. Divide your strategies into two lists.

- Strategies that cause a direct reduction (e.g. riding a bike instead of travelling in a car that uses fossil fuels)

- Strategies that cause an indirect reduction (e.g. buying products made using sustainable materials and processes)

4 Communicate

Use your list of strategies to construct a simple infographic to demonstrate the positive impacts of some of your carbon reduction strategies. The infographic on recycling shown here might give you some ideas. Remember to include:

- concise summaries of the facts

- diagrams and illustrations

- numbers and figures to support your arguments

- connections between cause and effect, using phrases such as, 'By doing X, you can help Y.' and 'Reducing A will cut down B.'

- positive language to encourage people to change

- solutions to problems.

RECYCLING AND COMPOSTING HELPS SAVE NATURAL RESOURCES.

RECYCLING AND COMPOSTING 87 MILLION TONS of MSW...

SAVED MORE THAN 1.1 QUADRILLION BTU OF ENERGY.

THAT'S THE SAME AMOUNT

OF ENERGY CONSUMED BY ALMOST 10 MILLION U.S. HOUSEHOLDS IN A YEAR.

EVERY TON OF PAPER RECYCLED

CAN SAVE THE ENERGY EQUIVALENT

OF 165 GALLONS OF GASOLINE.

RECYCLING 1 TON OF ALUMINUM CANS CONSERVES

EQUIVALENT TO 26 BARRELS OF OIL

OR 1,665 GALLONS OF GASOLINE.

OVER 153 MILLION BTUs.

MAKE A DIFFERENCE TODAY!

If we all take **small steps every day** to reduce the amount of waste we produce, **we can help protect our planet** for generations to come.

For more information, visit **www.epa.gov/recycle**

This infographic is based on data from EPA's 2012 MSW Characterization Report. For more information, see http://go.usa.gov/3-PzY.

Check out http://1.usa.gov/eswiafaq for the full infographic.

Energy transfer: sound, light and heat

7.1 Models for energy transfer (p. 216)

Scientific models can be applied to energy transfer contexts to help us understand many phenomena.

7.2 Types of waves (p. 218)

Waves can be classified into different types based on how they transfer energy.

7.3 Extension: Measuring the features of waves (p. 222)

Waves have different measurable features that are used to describe them.

7.4 Extension: More wave calculations (p. 228)

Mathematical relationships can be used to calculate features of waves.

7.5 Sound waves and hearing (p. 231)

Sound travels as longitudinal waves through different mediums, and the human ear has many components that interpret these waves as sounds.

7.6 Light waves and the electromagnetic spectrum (p. 236)

Light is one of many types of radiation that make up the electromagnetic spectrum.

7.7 Extension: The behaviour of light (p. 240)

When light meets new mediums, it undergoes processes such as reflection and refraction.

7.8 Thermal energy and temperature (p. 244)

Temperature is a measurement of kinetic energy and indicates changes in thermal energy.

7.9 Heat transfer (p. 246)

Thermal energy can be transferred as heat by three different methods: conduction, convection and radiation.

7.10 FIRST NATIONS SCIENCE CONTEXTS: First Nations Australians' traditional musical, hunting and communication instruments (p. 252)

Investigate the materials used to transfer sound energy in Australian First Nations Peoples' instruments.

7.11 SCIENCE AS A HUMAN ENDEAVOUR: The development of the cochlear implant (p. 255)

The cochlear implant was developed to provide hearing to people with profound hearing loss.

7.12 SCIENCE INVESTIGATIONS: Evaluating scientific models (p. 257)

1 Modelling waves with slinkys
2 Using a simulation model to investigate transverse waves
3 Speed of sound waves

BIG SCIENCE CHALLENGE #7

▲ **FIGURE 7.0.1** Sound barriers are often installed on busy roads.

Have you ever considered sound to be a type of pollution? Although it's invisible, noise pollution can pose a risk to living organisms, including humans. Traffic and construction noise can cause hearing loss and high blood pressure for humans. Noise pollution from ships can interfere with how dolphins and whales use sound to find their way through the ocean.

On major highways, engineered acoustic barriers are often installed to protect residents and nearby wildlife from the sounds of passing cars and trucks.

▶ Do you think there are times when the noise *you* produce might be harmful to other people or animals?

▶ Can you think of some other examples of ways we can reduce the negative effects of sound?

#7 SCIENCE CHALLENGE ACCEPTED!

At the end of this chapter, you can complete the Big Science Challenge Project #7. You can use the information you learn in this chapter to complete the project.

Assessments
- Prior knowledge quiz
- Chapter review questions
- End-of-chapter test
- Portfolio assessment task: Science investigation

Videos
- Science skills in a minute: Evaluating models **(7.12)**
- Video activities: What is sound? **(7.5)**; Electromagnetic spectrum **(7.6)**; What is light? **(7.6)**; Manipulating light **(7.7)**; The race for absolute zero **(7.8)**; Heat transport **(7.9)**

Science skills resources
- Science skills in practice: Evaluating scientific models **(7.12)**
- Extra science investigations: Reflection and refraction **(7.7)**; Comparing thermal conductivity **(7.9)** Slinky extension **(7.12)**

Interactive resources
- Simulation: Waves introduction **(7.2)**; Molecules and light **(7.7)**; Bending light **(7.7)**; Wave on a string **(7.12)**
- Label: Features of waves **(7.3)**; Parts of the ear **(7.5)**
- Crossword: Types of waves **(7.2)**

Nelson MindTap

To access these resources and many more, visit:
cengage.com.au/nelsonmindtap

7.1 Models for energy transfer

Interactive resource
Drag and drop: Particle or wave model?

GET THINKING

Look around your science classroom. Are there different representations, such as diagrams or physical models, that are used to help you understand complex science concepts? Make a list of as many as you can.

Scientific models

A **model** is a representation that describes, simplifies or provides an explanation of the workings, structure or relationships within a particular **phenomenon**, object or idea. Models can be diagrams, physical models, computer programs, simulations, analogies or complicated mathematical equations. A very simple example is the 'cloud in a jar' model of the formation of clouds, shown in Figure 7.1.1.

model
a conceptual, physical or mathematical representation of an idea, phenomenon or process

phenomenon
something that is observed to exist or occur (plural: phenomena)

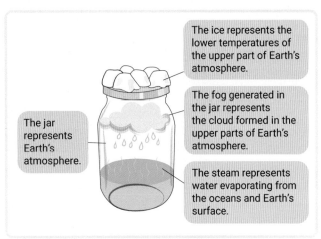

The ice represents the lower temperatures of the upper part of Earth's atmosphere.

The fog generated in the jar represents the cloud formed in the upper parts of Earth's atmosphere.

The jar represents Earth's atmosphere.

The steam represents water evaporating from the oceans and Earth's surface.

▲ **FIGURE 7.1.1** A scientific model for cloud formation: 'cloud in a jar'. Ice is put on top of a jar containing boiling water.

Models are often developed and refined over time as more evidence and observations about an idea are collected. Models are never perfect at representing phenomena. Some models can get very close to the real thing, but most have their limitations. For example, the 'cloud in a jar' model helps us to understand the relationships between temperature differences and how water changes phase in the sky, but it doesn't help us to understand cloud shape or movement through the atmosphere.

Models for energy transfer

Energy is a concept that, in many ways, scientists still struggle to understand. **Energy** is defined as the ability to do work. We can also think of energy as something that makes things happen. Most forms of energy can be classified as either potential energy or kinetic energy. Some examples of energy are thermal energy, sound energy, light energy, chemical energy, nuclear energy and electrical energy. Energy can move from one place to another (be transferred) in many different ways. Scientists use different models to understand the process of **energy transfer**. The most common models for energy transfer are the **wave** and **particle** models.

energy
the ability to do work; measured in joules (J), kilojoules (kJ) or megajoules (MJ)

energy transfer
the movement of energy without changing its type

wave
a disturbance of energy that moves through matter or space

particle
a small, localised piece of matter, usually modelled to be spherical

The wave model

The wave model explains how energy can travel as a wave through space without carrying matter along the way. Waves can carry energy through space to objects, where **energy transformation** takes place – the energy is converted into other forms of energy. For example, the Sun's energy travels through space to plants, where it is transformed through photosynthesis into chemical energy for food.

A wave is easy to visualise by imagining a bird floating in water as a wave passes (Figure 7.1.2). While it might appear that water is travelling along with the wave, the water molecules, and the bird, are really just bobbing up and down as the energy of the wave moves through the water.

The particle model

A particle can be an atom, a molecule or even a tennis ball, depending on the context. The particle model explains how energy can be transferred from particle to particle because of the particles' movements. The particle model is based on the idea that particles contain energy that they can transfer to other particles or places. Particles can do this by:

- passing energy between one another through direct contact
- moving through space and carrying energy with them, transferring the energy to a new space.

In both of these cases, the energy is moving but staying in the same form, making them both types of energy transfer (Figure 7.1.3).

The duality of light

Throughout history, scientists have performed experiments that help us to understand what light is and how it works. Many experiments suggest that light is a wave; applying a wave model allows us to explain light's behaviour and make predictions. However, other experiments suggest that light is made up of individual particles moving through space, which can be explained by the particle model but not by a wave model. There is no 'right' or 'wrong' answer about whether light is a wave or a particle. Both of these ideas are models that we use to try to understand light. And both work in different situations! But, in Year 9, we will model light as a wave.

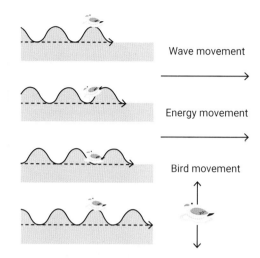

▲ **FIGURE 7.1.2** A bird floating on the water moves up and down as the energy moves left to right, perpendicular to the oscillation of the particles when a wave moves through the water.

energy transformation a movement of energy where it changes from one form into another

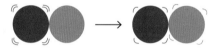

Energy is passed from the left particle to the right particle.

Energy is passed from left to right as the particle moves.

▲ **FIGURE 7.1.3** Particles can transfer energy between each other or through space by moving.

 7.1 LEARNING CHECK

1 **Define**:
   a scientific model.   b phenomenon.
2 Give three examples of scientific models you have learned about or used before in science to help you understand something.
3 Take three photos around your house or classroom of places where you can observe energy moving by sound, light or heat. Use these pictures to draw sketches of the situations.

7.2 Types of waves

✓ describe how mechanical and electromagnetic waves transfer energy

✓ compare transverse and longitudinal mechanical waves by identifying and describing their features.

GET THINKING

Imagine a Mexican wave in a stadium at a football match. If you can't imagine this, look up a video of it on the Internet. Why do you think it is called a 'wave'? What does it have in common with other waves you know about?

Types of waves

There are two main types of waves: mechanical and electromagnetic. They are classified according to whether they can or cannot transmit energy through a **vacuum**.

Mechanical waves transfer energy through matter by the vibrations of particles (Figure 7.2.1). The matter they move through is referred to as the **medium**. Mechanical waves cannot travel through a vacuum where there are no particles, such as in outer space. Mechanical waves include sound waves, ocean waves, seismic waves and the waves you make when you flick a piece of rope or a string.

vacuum
a space where there is no matter (no particles)

mechanical wave
a wave that passes energy through space between particles

medium
the matter or substance through which a wave passes (made from particles)

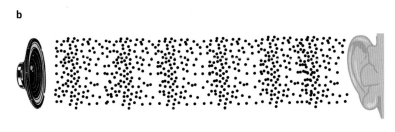

▲ **FIGURE 7.2.1** Some examples of mechanical waves: **(a)** a wave moving through water; **(b)** a sound wave.

electromagnetic wave
a wave that transfers energy through space by electric and magnetic fields

Electromagnetic waves are waves that transfer energy by electric and magnetic fields (Figure 7.2.2). Light, infrared (heat) radiation, X-rays and ultraviolet (UV) rays are all electromagnetic waves. Electromagnetic waves do not use particle movement to transfer energy, and can travel through a vacuum. This is why light and heat are able to travel through outer space from the Sun to Earth. Electromagnetic waves are covered in more detail in Module 7.5.

Interactive resources
Simulation: Waves introduction

Crossword: Types of waves

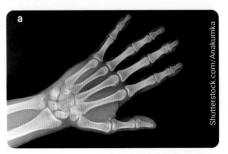

▲ **FIGURE 7.2.2** Some examples of electromagnetic waves: **(a)** X-rays can pass through your body; **(b)** microwaves are used in speed detection radar guns.

Mechanical waves

There are two types of mechanical waves: transverse waves and longitudinal waves. Mechanical waves are classified according to the direction that the particles move compared with the direction of the energy transfer.

Transverse waves transfer energy by the **oscillation** of particles in a direction that is perpendicular (at right angles) to the direction that energy is travelling, like a wave in the ocean. You can see a transverse wave in Figure 7.2.3. An oscillation is a repeated vibration back and forth or up and down. The particles in a transverse wave vibrate up and down as the wave passes from left to right. On a transverse wave, the highest points are known as **crests** and the lowest points are known as **troughs**.

transverse wave
a mechanical wave in which particles oscillate at right angles to the direction in which the wave is travelling

oscillation
movement back and forth in a regular rhythm or pattern

crest
the highest point on a transverse wave

trough
the lowest point on a transverse wave

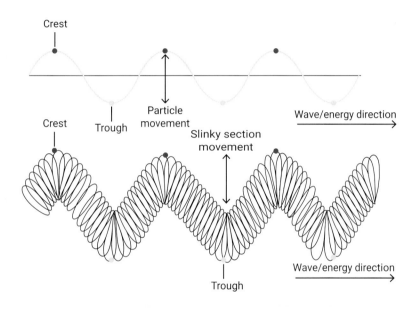

▲ **FIGURE 7.2.3**   In a transverse wave, the particles oscillate at right angles to the direction in which the wave is moving. You can model this type of wave with a slinky.

longitudinal wave
a mechanical wave in which particles oscillate in the same direction in which the wave is travelling

compression
a high-pressure region of a longitudinal mechanical wave where particles are pushed close together

rarefaction
a low-pressure region of a longitudinal mechanical wave where particles are spread apart

Longitudinal waves are mechanical waves that transfer energy by the oscillation of particles in a direction parallel to the direction the energy is travelling, like a sound wave. You can see a longitudinal wave in Figure 7.2.4. This means that if the wave is travelling left to right, the particles in the wave also vibrate left and right as the wave passes. The particles are compressed together and spread apart as the wave passes. On a longitudinal wave, there are zones of **compression** where the particles are closest together and zones of **rarefaction** where the particles are most spread out.

You will model waves using slinkys in Module 7.12 Investigation 1.

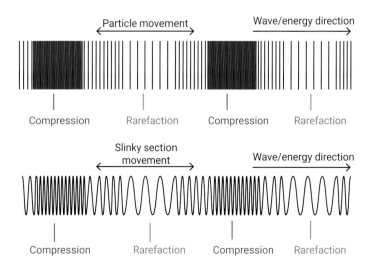

▲ **FIGURE 7.2.4**   In a longitudinal wave, the particle oscillates in the same direction as the wave. You can model this type of wave with a slinky.

Features of mechanical waves

Mechanical waves can differ from each other by some main features: **amplitude**, **wavelength**, **frequency**, **period** and **speed**.

▼ TABLE 7.2.1 A summary of wave features

Feature	Definition	Symbol	Units
Amplitude	• A measure of the extent of particle movement • The greater the energy of a wave, the more particle movement and the larger the amplitude	A	metres (m)
Wavelength	• A measurement of how long each wave is • Can be measured from any point on a wave to the same point on the next wave	λ (Greek letter lambda)	metres (m)
Frequency	• A measure of how many full waves pass by a point each second	f	hertz (Hz) or per second (/s)
Period	• The time it takes for one full wave to pass a particular point	T	seconds
Speed	• How far a wave travels in a certain time • Depends on the medium that a mechanical wave is travelling through	v	m/s

amplitude
the maximum distance that a particle moves when a mechanical wave passes through it

wavelength
the length of one full wave; the distance between two crests or two troughs; measured in metres (m)

frequency
the number of waves produced, or passing a point per unit of time; measured in hertz (Hz)

period
the time taken for one full wave to pass a particular point

speed
a measure of how fast a wave is travelling; a measure of the distance covered by the wave per unit of time; measured in metres per second (m/s)

In Module 7.3, you will see how these features can be measured and the formulas used to calculate them in some worked examples.

Features of electromagnetic waves

Many of the features of mechanical waves can also be applied to electromagnetic waves. Electromagnetic waves are sometimes classified as a type of transverse wave. Although they don't have particle movement perpendicular (at right angles) to the wave direction, electromagnetic waves have electric and magnetic fields that oscillate perpendicular to the wave's direction (Figure 7.2.5).

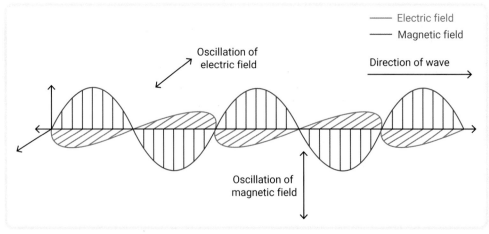

▲ **FIGURE 7.2.5** Electric and magnetic fields oscillate at 90° (perpendicular) to the direction of the electromagnetic wave.

9780170472838

Like mechanical transverse waves, electromagnetic waves have crests and troughs. This means the amplitude, wavelength and period of an electromagnetic wave can still be measured. It also means frequency and speed measurements apply to electromagnetic waves. The differences in frequency and wavelength between different electromagnetic waves are the features that allow them to be classified into different groups such as light, X-rays and radio waves. We will cover more on this in Module 7.6.

7.2 LEARNING CHECK

1 **Distinguish** between mechanical waves and electromagnetic waves.

2 **Define:**
   a crest.
   b trough.
   c compression.
   d rarefaction.

3 **Copy and complete** the table to summarise wave features.

Feature	Definition	Symbol	Units
Wavelength			
Frequency			
Speed			
Amplitude			

4 a **Copy and complete** the following sentence about transverse waves.
     The _____ in a transverse wave results in the movement of particles in a _____ direction compared with the _____ of the wave's motion.

   b **Write** a similar sentence to the one in part **a** for longitudinal waves that describes the direction of particle movement.

5 **Compare** longitudinal waves and transverse waves.

BY THE END OF THIS MODULE, YOU WILL BE ABLE TO:

✓ use wave graphs to identify the amplitude, wavelength and period of a wave

✓ use the formula for frequency and speed to calculate these features of a wave.

Interactive resource
Label: Features of waves

GET THINKING

Can you think of examples of when we might need to measure the features of waves, such as a wave's speed, frequency or amplitude? Why do you think this information can be helpful?

Measuring amplitude

The amplitude of a transverse wave measures half the distance from a crest to the trough. It can be measured from the middle of the wave (the rest axis) to a crest or to a trough. The amplitude for a longitudinal wave can be measured by how far particles have travelled when compressed together at a compression point or how far they have travelled apart at a rarefaction point. This is usually quite hard to measure.

Measuring wavelength

Wavelength can be measured from any point on a wave to the same point on the next wave. The easiest ways to measure a wavelength for each type of wave are:

- transverse wave: the distance from crest to crest or from trough to trough

- longitudinal wave: the distance from compression to compression or from rarefaction to rarefaction.

This can be done when observing waves in different mediums such as water (see Figure 7.3.1) or using distance wave graphs, as shown in Figure 7.3.2.

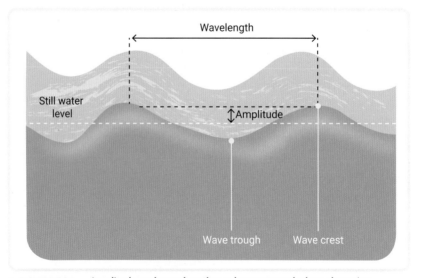

▲ **FIGURE 7.3.1** Amplitude and wavelength can be measured when observing waves moving through the water.

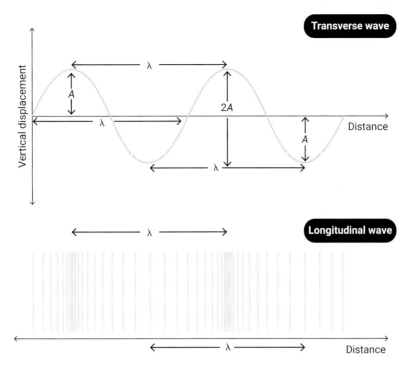

▲ **FIGURE 7.3.2** Graphs of vertical displacement against distance for transverse and longitudinal waves. *A* is amplitude and λ is wavelength.

Measuring frequency

The frequency (*f*) can be determined by counting how many full waves pass by a single point in a certain amount of time:

$$f = \frac{\text{number of full waves passed in time period}}{\text{length of time}}$$

where length of time is measured in seconds (s) and frequency is measured in hertz (Hz).

▲ **FIGURE 7.3.3** You can work out the frequency of waves by counting how many full waves pass by a particular point in a given time.

WORKED EXAMPLE 7.3.1

During a 4-minute period, 48 water waves pass a point in a river. Calculate the frequency of the waves.

THINKING PROCESS	WORKING
Step 1: Identify the known and unknown variables.	Length of time = 4 minutes Number of waves = 48 Frequency (f) = ?
Step 2: Ensure that all known variables are given in appropriate units. Convert if required.	Length of time = 4 minutes = 4 × 60 seconds = 240 s
Step 3: Identify the appropriate relationship.	$f = \dfrac{\text{number of full waves passed in time period}}{\text{length of time period}}$
Step 4: Substitute known values.	$f = \dfrac{48 \text{ waves}}{240 \text{ s}}$
Step 5: Rearrange, using algebra if necessary, and solve.	$f = 0.2 \text{ Hz}$

Measuring period

The period (T) is determined by timing how long it takes for one wave to pass or counting the number of full waves that pass by a single point in a certain amount of time:

$$T = \frac{\text{length of time period}}{\text{number of full waves passed in time period}}$$

where length of time and period are both measured in seconds (s).

WORKED EXAMPLE 7.3.2

During a 4-minute period, 48 water waves pass a point in a river. Calculate the period of each wave.

THINKING PROCESS	WORKING
Step 1: Identify the known and unknown variables.	Length of time = 4 minutes Number of waves = 48 Period (T) = ?
Step 2: Ensure that all known variables are given in appropriate units. Convert if required.	Length of time = 4 minutes = 4 × 60 seconds = 240 s
Step 3: Identify the appropriate relationship.	$T = \dfrac{\text{length of time period}}{\text{number of full waves passed in time period}}$
Step 4: Substitute known values.	$T = \dfrac{240 \text{ s}}{48 \text{ waves}}$
Step 5: Rearrange, using algebra if necessary, and solve.	$T = 5 \text{ s}$

9780170472838

If a time wave graph is given where time is given on the *x*-axis instead of distance (as in Figure 7.3.4), then the period can be measured directly as the time of one full wave.

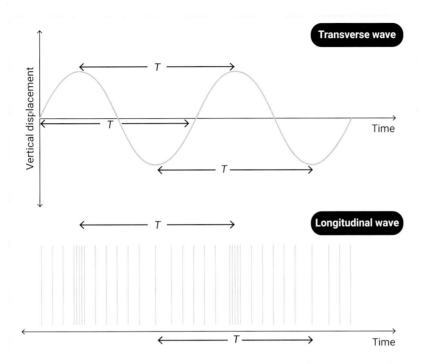

▲ **FIGURE 7.3.4**   Graphs of vertical displacement against time for transverse and longitudinal waves. *T* is period.

Measuring speed

Wave speed (*v*) can be calculated by measuring how far a wave travels in a certain time:

$$v = \frac{\text{distance travelled by wave}}{\text{length of time}}$$

where distance travelled is measured in metres (m), time is measured in seconds (s) and speed is measured in metres per second (m/s).

◀ **FIGURE 7.3.5**   Radar guns determine a car's speed by sending out and measuring the 'echo' of radio waves, which travel through air at a constant speed.

A person stands 170 m from a vertical cliff and shouts. After 1 s, the person hears their shout return to them. Calculate the speed of the sound wave.

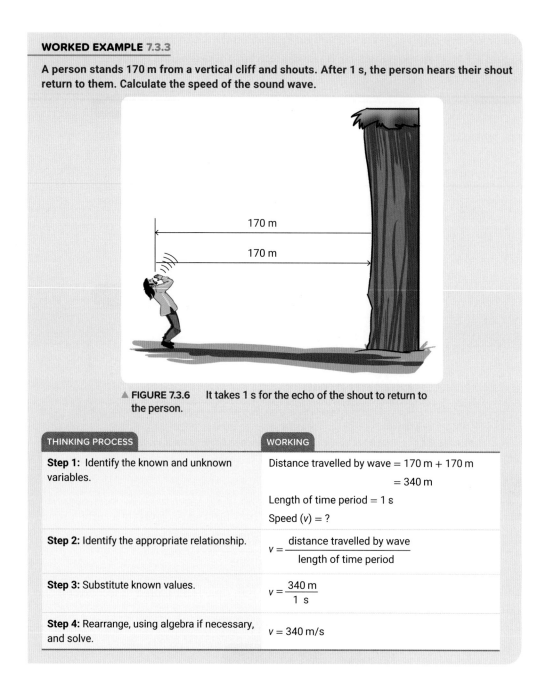

▲ **FIGURE 7.3.6**   It takes 1 s for the echo of the shout to return to the person.

THINKING PROCESS	WORKING
Step 1: Identify the known and unknown variables.	Distance travelled by wave = 170 m + 170 m = 340 m Length of time period = 1 s Speed (v) = ?
Step 2: Identify the appropriate relationship.	$v = \dfrac{\text{distance travelled by wave}}{\text{length of time period}}$
Step 3: Substitute known values.	$v = \dfrac{340 \text{ m}}{1 \text{ s}}$
Step 4: Rearrange, using algebra if necessary, and solve.	$v = 340$ m/s

You will use a simulation model to look at wave features in Module 7.12 Investigation 2.

9780170472838

1 Using the graph in Figure 7.3.7, **identify** the:

  a   wavelength.

  b   amplitude.

  c   number of full wavelengths shown.

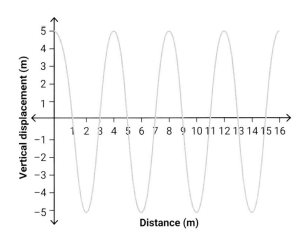

▲ **FIGURE 7.3.7**   A graph of vertical displacement against distance for a transverse wave

2 Using the graph in Figure 7.3.8:

  a   **identify** the period of the wave.

  b   **calculate** the frequency of the wave.

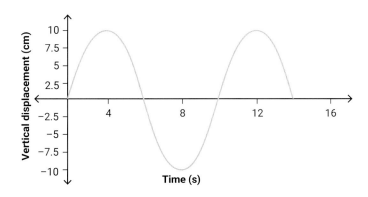

▲ **FIGURE 7.3.8**   A graph of vertical displacement against time for a transverse wave

BY THE END OF THIS MODULE, YOU WILL BE ABLE TO:
- ✓ calculate the frequency of a wave when given the period and vice versa
- ✓ calculate the frequency, wavelength and speed of a wave using the wave equation.

Frequency and period calculations

Period and frequency are reciprocals of one another. A high frequency means a low period and a low frequency means a high period. The period can be calculated from the frequency and vice versa.

$$\text{Period} = \frac{1}{\text{frequency}} \quad \text{and frequency} = \frac{1}{\text{period}}$$

$$T = \frac{1}{f} \quad \text{and} \quad f = \frac{1}{T}$$

where period is measured in seconds (s) and frequency is measured in hertz (Hz).

WORKED EXAMPLE 7.4.1

A sound wave takes 0.02 s to pass a particular point. Calculate the frequency of the sound wave.

THINKING PROCESS	WORKING
Step 1: Identify the known and unknown variables.	$T = 0.02$ s $f = ?$
Step 2: Identify the appropriate relationship.	$f = \dfrac{1}{T}$
Step 3: Substitute known values.	$f = \dfrac{1}{0.02 \text{ s}}$
Step 4: Rearrange, using algebra if necessary, and solve.	$f = 50$ Hz

WORKED EXAMPLE 7.4.2

A wave machine in a wave pool generates waves at a frequency of 0.5 Hz. Calculate the period of each wave.

THINKING PROCESS	WORKING
Step 1: Identify the known and unknown variables.	$f = 0.05$ Hz $T = ?$
Step 2: Identify the appropriate relationship.	$T = \dfrac{1}{f}$
Step 3: Substitute known values.	$T = \dfrac{1}{0.5 \text{ Hz}}$
Step 4: Rearrange, using algebra if necessary, and solve.	$T = 2$ s

The wave equation

Another important relationship that can be used to analyse waves is the **wave equation**. The wave equation is a mathematical relationship between the speed, wavelength and frequency of a wave:

wave equation
a mathematical relationship between the speed, wavelength and frequency of a wave

$$\text{Speed} = \text{frequency} \times \text{wavelength}$$

$$v = f \times \lambda$$

where speed is measured in metres per second (m/s), wavelength is measured in metres (m) and frequency is measured in hertz (Hz).

WORKED EXAMPLE 7.4.3

A wave has a frequency of 50 Hz and a wavelength of 10 m. Calculate the speed of the wave.

THINKING PROCESS	WORKING
Step 1: Identify the known and unknown variables.	$f = 50\,\text{Hz}$ $\lambda = 10\,\text{m}$ $v = ?$
Step 2: Identify the appropriate relationship.	$v = f \times \lambda$
Step 3: Substitute known values.	$v = 50\,\text{Hz} \times 10\,\text{m}$
Step 4: Rearrange, using algebra if necessary, and solve.	$v = 500\,\text{m/s}$

WORKED EXAMPLE 7.4.4

A wave rolling through a harbour is travelling at 3 m/s and has a wavelength of 0.5 m. Calculate the frequency of the wave.

THINKING PROCESS	WORKING
Step 1: Identify the known and unknown variables.	$v = 3\,\text{m/s}$ $\lambda = 0.5\,\text{m}$ $f = ?$
Step 2: Identify the appropriate relationship.	$v = f \times \lambda$
Step 3: Substitute known values.	$3\,\text{m/s} = f \times 0.5\,\text{m}$
Step 4: Rearrange, using algebra if necessary, and solve.	$f = \dfrac{3\,\text{m/s}}{0.5\,\text{m}}$ $= 6\,\text{Hz}$

A wave made within a skipping rope is travelling at 8 m/s and 16 waves are produced every second. Calculate the wavelength of the wave.

THINKING PROCESS	WORKING
Step 1: Identify the known and unknown variables.	$v = 8 \text{ m/s}$ $f = 16 \text{ Hz}$ $\lambda = ?$
Step 2: Identify the appropriate relationship.	$v = f \times \lambda$
Step 3: Substitute known values.	$8 \text{ m/s} = 16 \text{ Hz} \times \lambda$
Step 4: Rearrange, using algebra if necessary, and solve.	$\lambda = \dfrac{8 \text{ m/s}}{16 \text{ Hz}}$ $= 0.5 \text{ m}$

7.4 LEARNING CHECK

1 A tap is dripping into a bowl of water. Drops of water produce water waves from the point where the drops hit the surface of the water. If the frequency of drops from the tap is 5 Hz, **calculate** the period of each wave produced.

2 If waves at the beach are 4 m long and two waves reach the shoreline every 10 s, **calculate** the:

a frequency of the waves.

b speed of the waves.

▲ FIGURE 7.4.1 Dripping water making waves along the water's surface

3 Red light travels at a speed of 300 000 000 m/s and has a wavelength of 0.000 0007 m. **Calculate** the frequency of the waves of red light.

▲ FIGURE 7.4.2 Red light travels at the speed of light but has a different frequency and wavelength from other colours of light. Red light has the longest wavelength of all the colours so it can be seen from the greatest distances.

9780170472838

BY THE END OF THIS MODULE, YOU WILL BE ABLE TO:
- ✓ describe how sound energy is transferred
- ✓ construct representations of sound waves and identify key features, including amplitude and frequency
- ✓ compare the volume and pitch of different sound waves
- ✓ describe the structure and function of the human ear and explain how we can hear.

GET THINKING

List two things you already know about sound. Write down two things about sound that you don't know how to explain but would like to learn more about.

Video activity
What is sound?

Interactive resource
Label: Parts of the ear

Making and moving sound

Sounds are caused by the transfer of kinetic energy through a medium by longitudinal waves. When you clap your hands, the kinetic energy of your hands is transferred to the air particles around your hands. These vibrations pass through the air to your ears where the particle vibrations are turned into sounds by your ears and brain. A similar thing happens when you bang a drum and the vibration of the drum skin passes a repeating wave of compressions through the air, as shown in Figure 7.5.1.

When you hear sounds under water, the kinetic energy is passed between water molecules to the air particles trapped in your ear canal and received by your ears in the same way. Because sound is transferred between particles, it requires a medium; both air and water provide these particles. Sound cannot travel through a vacuum because it contains no particles.

Figure 7.5.2 shows the air particles in a silent room. A sound wave then travels through the air, and the air particles compress together and spread apart in a repeated pattern.

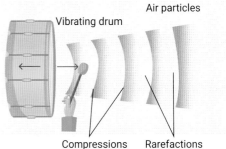

▲ **FIGURE 7.5.1** When a drum has been struck, the drum skin's vibrations cause the air to vibrate, and the air vibrations pass sound energy between them as a longitudinal wave.

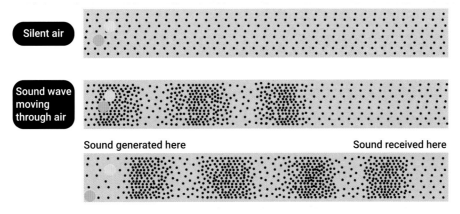

▲ **FIGURE 7.5.2** Air particles in a silent room and then when a sound wave moves through the air. The enlarged coloured dots are two air particles highlighted to show their movement as the wave passes.

The particles oscillate backwards and forwards as the sound waves move past them, forming a longitudinal mechanical wave. When this happens to the air particles inside your ears, you hear it as a sound.

When sound waves hit a hard boundary, the energy is not always absorbed by the material; instead, the sound waves may reflect off the surface and travel back in the opposite direction. This is what happens when you hear an echo (Figure 7.5.3).

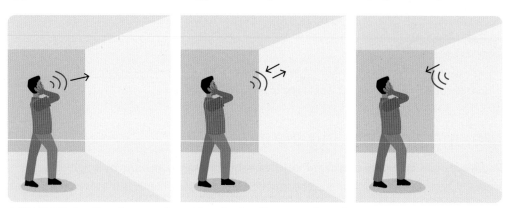

▲ **FIGURE 7.5.3**   Sound waves rebound from hard surfaces, creating echoes.

Features of sound waves

Sound waves can vary in amplitude, frequency and wavelength.

The amplitude for a sound wave measures how far particles in a medium have travelled when compressed together at a compression point or how far they have travelled apart at a rarefaction point. When more energy is being transferred, the particles are compressed and spread further and so the amplitude of the wave is greater.

volume
the measure of how loud a sound is; measured in decibels (dB)

Sound waves carrying more energy are louder than other waves. The amplitude of a wave is a direct indicator of the **volume** of the wave. Figure 7.5.4 shows a low-volume and a high-volume sound wave. A low-volume sound wave could be if you whistled a high note very softly. A high-volume sound wave might be the wave when the same note is whistled very loudly.

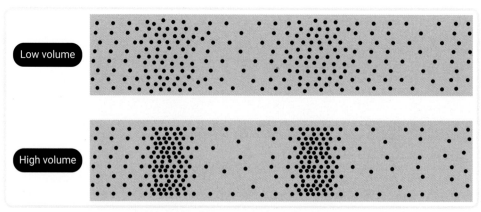

▲ **FIGURE 7.5.4**   Two sound waves of the same pitch moving through air, one low volume and one high volume

9780170472838

The frequency of a sound wave matches the frequency of the vibrations that cause the sound wave to travel through the medium. A higher frequency sound is heard as a higher note or higher **pitch**.

pitch
the degree of frequency (high or low) of a sound or note

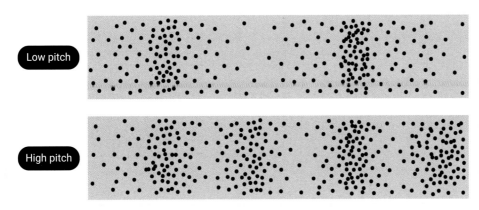

▲ **FIGURE 7.5.5** Two sound waves of the same volume moving through air, one low pitch and one high pitch

The wavelength of a sound wave can vary and will depend on how fast the wave is travelling and the frequency of the sound.

Visualising sound waves

Making measurements of sounds and comparing different sound waves can be quite difficult when the sound waves are visualised as they are in Figures 7.5.4 and 7.5.5. You might have already seen sound waves visualised as more like transverse waves going up and down, as in Figure 7.5.6.

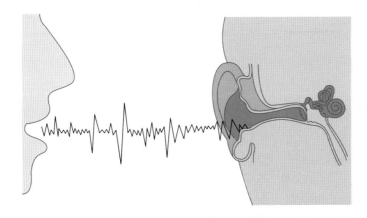

▲ **FIGURE 7.5.6** A more common visualisation of a sound wave, looking like a transverse wave

Sometimes it is easier to graph sound waves on a compression graph. Figure 7.5.7 shows how a sound wave can be displayed on a compression graph by plotting how compressed the particles are at each point along the wave. This makes the amplitude and wavelength much easier to see and visualise for different waves. Sound waves of different volume and pitch are shown in Figure 7.5.8.

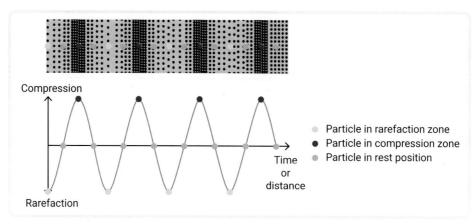

▲ FIGURE 7.5.7   Converting a sound wave to a compression graph makes it easy to visualise and identify key features.

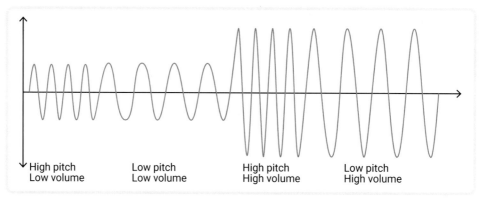

▲ FIGURE 7.5.8   Sound waves of varying pitch (frequency) and volume (amplitude)

How we hear sound

The ear is made up of many smaller structures that allow humans to hear sounds. Figure 7.5.9 shows the various structures within the ear and how these structures allow hearing.

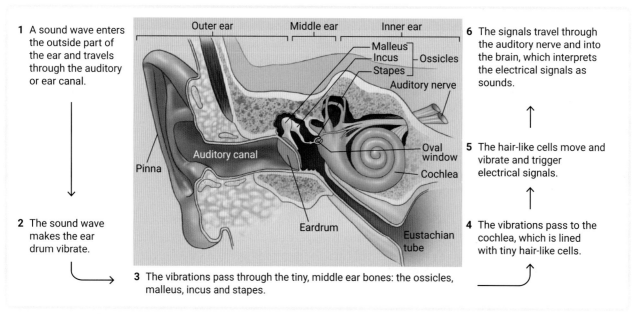

1 A sound wave enters the outside part of the ear and travels through the auditory or ear canal.

2 The sound wave makes the ear drum vibrate.

3 The vibrations pass through the tiny, middle ear bones: the ossicles, malleus, incus and stapes.

4 The vibrations pass to the cochlea, which is lined with tiny hair-like cells.

5 The hair-like cells move and vibrate and trigger electrical signals.

6 The signals travel through the auditory nerve and into the brain, which interprets the electrical signals as sounds.

▲ FIGURE 7.5.9   The structure of the human ear and the process of hearing

Observing sound

1 Put different amounts of water in three bottles.

2 Blow over the top of the bottles to produce a sound.

3 Write a conclusion about whether higher frequency sound waves are produced in fuller or emptier bottles. Give reasons for your answer.

 **7.5 LEARNING CHECK**

1 a **Draw** a compression graph for a sound wave similar to the one in Figure 7.5.10, with an amplitude of 2 units and wavelength of 4 units

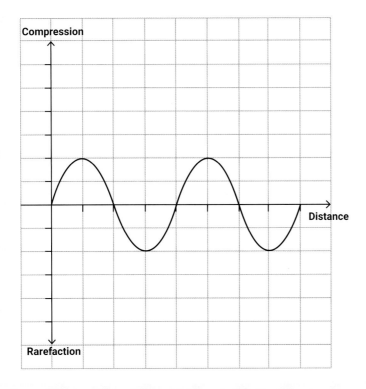

▲ **FIGURE 7.5.10**   A sound wave with an amplitude of 2 units and wavelength of 4 units

   b Using a second colour, **draw** a sound wave over the top that is twice as loud as the wave in part **a**.

   c Using a third colour, **draw** a sound wave over the top that has a lower pitch than the wave in part **a**.

   d Using a fourth colour, **draw** a sound wave that is half as loud and that has a higher pitch than the wave in part **a**.

2 As people get older, the tiny hair-like cells inside the cochlea can become damaged. **Infer** what this might mean for the hearing of people as they age.

✓ recognise that visible light is a type of electromagnetic wave and each type has different wave features

✓ explain light phenomena and behaviour using the wave model, including how we see different colours.

Video activities
Electromagnetic
spectrum

What is light?

GET THINKING

Skim read this module and take note of the headings, pictures and diagrams. Then make a list of three things you think are going to be covered in this module that you are interested in learning more about.

electromagnetic radiation
energy that travels in electromagnetic waves

electromagnetic spectrum
the range of electromagnetic waves arranged in sequential order from high-energy gamma rays to low-energy radio waves

The electromagnetic spectrum

Electromagnetic waves do not require particles to transfer energy and so can travel through a vacuum. Electromagnetic waves are sometimes referred to as **electromagnetic radiation** (or just radiation). Electromagnetic waves carry different amounts of energy, depending on their frequency and wavelength. Scientists have classified the types of radiation into seven main groups on the **electromagnetic spectrum**. Figure 7.6.1 shows the electromagnetic spectrum, including gamma radiation, X-ray radiation, ultraviolet (UV) radiation, visible light, infrared (IR) radiation, microwave radiation and radio wave radiation.

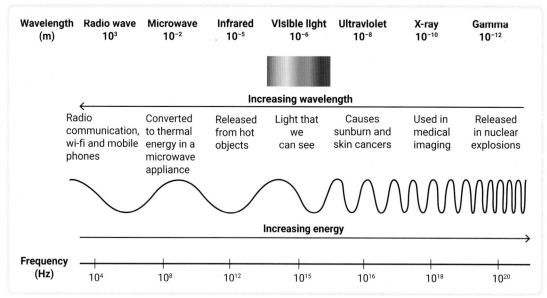

▲ **FIGURE 7.6.1**   The different electromagnetic waves of the electromagnetic spectrum

The waves of the electromagnetic spectrum are an important part of our lives. We use light to see, infrared radiation from the Sun heats Earth, and we warm food with microwaves. Without electromagnetic waves, there would be no X-ray machines, wireless Internet or mobile phones. Figure 7.6.1 provides some more information about each type of electromagnetic radiation.

A **spectrum** is a scale. The electromagnetic spectrum shows the specific frequency range and wavelength range of each type of radiation. This determines which group an electromagnetic wave belongs to. The spectrum also orders types of radiation by how much energy they carry. Radio waves have the lowest frequency and longest wavelength, and they carry the least amount of energy.

Gamma rays have the highest frequency and shortest wavelength, and they carry the most energy. The higher the frequency of electromagnetic radiation, the more energy carried by the wave.

Electromagnetic waves can have different frequencies and wavelengths, but they all have the same speed when travelling through the same medium. In a vacuum, the speed of electromagnetic radiation is 300 000 km/s or 3×10^8 m/s. This value is often referred to as the speed of light because light is one type of electromagnetic radiation.

Light: an electromagnetic wave

Visible light is the only part of the electromagnetic spectrum we can see. Humans cannot see radiation outside of the wavelength range of about 380–750 nm (nanometres) (3.8×10^{-7} m to 7.5×10^{-7} m). Other animals can detect different parts of the electromagnetic spectrum. For example, some insects and birds can see ultraviolet light.

Visible light can be many different colours depending on the wavelength. Scientists often refer to seven different colours of the visible light spectrum as ROYGBIV: red, orange, yellow, green, blue, indigo and violet. Violet light has the shortest wavelength and carries the most energy.

When our eyes receive all colours of the visible light spectrum at the same time, the light we see is white. The different colours in white light can be separated by passing it through a prism. Water droplets in the air can also act like a prism, splitting the white light from the Sun to form rainbows (Figure 7.6.2). Next time you see a rainbow, remember ROYGBIV and see if you can spot each colour!

▲ **FIGURE 7.6.2**   A rainbow forms when white light passes through water droplets and is split into different colours.

Where does light come from?

Light is produced when electromagnetic energy is released from somewhere and has a wavelength in the visible light section of the spectrum. Sometimes when objects get extremely hot and the radiation has more energy than infrared heat radiation, light is released. Flames are an example. The Sun is constantly undergoing nuclear reactions that release large amount of radiation, including visible light, UV light and infrared radiation. Some living things can produce their own light through a process of bioluminescence (Figure 7.6.3).

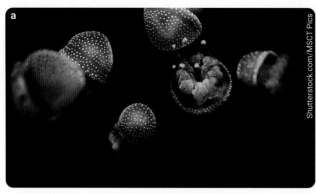
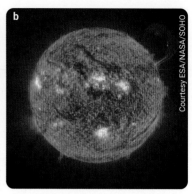

▲ **FIGURE 7.6.3** Light can be produced when energy is released such as **(a)** in a biological reaction (bioluminescence) or **(b)** in a nuclear reaction.

However, most light that we see does not come directly from its source. We can see objects because light reflects off them. When light (or any electromagnetic radiation) contacts a surface, three things can happen. The light can:

- be reflected off the surface
- be absorbed by the material of the surface
- pass through the surface (transmission).

Even though light is a wave, it is often visualised as an arrow or a ray in diagrams to show the direction and motion of the light wave. The **reflection**, **absorption** and transmission of light is shown in Figure 7.6.4.

Mirrors are surfaces that reflect all wavelengths of light. Black surfaces absorb most wavelengths of light, and transparent surfaces transmit light through them. Other surfaces reflect, absorb and transmit different waves of light depending on the wavelength. This is the reason we see different colours on different surfaces.

reflection
the process of a wave bouncing off a surface and returning in the opposite direction through the same medium

absorption
the process of taking in radiation and converting it to other types of energy

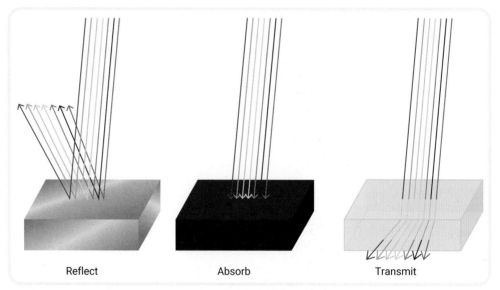

Reflect   Absorb   Transmit

▲ **FIGURE 7.6.4** When light contacts a surface, it is reflected, absorbed or transmitted.

9780170472838

Seeing colours

In a dark room where there is no light, everything is black. When a light is turned on or sunlight enters through a window, a red delicious apple in the room appears red to our eyes. White light containing all ROYGBIV colours shines onto the apple. Because of the chemicals in the apple skin, it absorbs orange, yellow, green, blue, indigo and violet (OYGBIV) light, but reflects red light off the surface (see Figure 7.6.5). The reflected light travels to our eyes and our brain interprets the colour of the object. Black surfaces are 'colourless' because they absorb all colours of light.

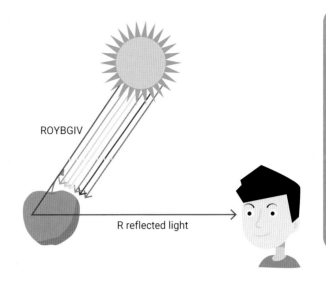

▲ FIGURE 7.6.5   We see different colours when certain colours of light are reflected from surfaces to our eyes.

Observing reflection and absorption

✩ ACTIVITY

Place a white piece of paper and a black piece of paper outside in sunlight. Hold your palm above each piece of paper and observe how bright your palm is. **Explain** the differences between the two pieces of paper in terms of reflection and absorption.

7.6 LEARNING CHECK

1 **Copy and complete** the following sentences about electromagnetic waves.

   a Waves that have short wavelengths have _____ frequencies and carry _____ energy. An example of this is a _____ wave.

   b Waves that have long wavelengths have _____ frequencies and carry _____ energy. An example of this is a _____ wave.

   c Electromagnetic waves that are visible to the human eye are known as visible light. They have wavelengths longer than _____ light but shorter than _____ radiation.

   d Compared to green light, blue light has a _____ frequency and _____ wavelength.

2 **Calculate** the frequency of ultraviolet light with wavelength of 0.000 000 350 m.

3 Look at the image to the right and answer the following questions.

   a **Explain** why leaves appear green in the daylight.

   b **Predict** what colour the leaves would appear if you shone a pure red light on them in a dark room. **Explain** your answer.

▲ FIGURE 7.6.6   Why do leaves appear green?

Chapter 7 | Energy transfer: sound, light and heat

Extension: The behaviour of light

BY THE END OF THIS MODULE, YOU WILL BE ABLE TO:

✓ describe how light waves reflect and refract

✓ describe the structure and function of the human eye and explain how we can see.

Video activity
Manipulating light

Interactive resources
Simulations:
Bending light

Molecules and light

Extra science investigation
Reflection and refraction

GET THINKING

Understanding the behaviour of light means that we can use light to help us do different things. Make a list of specific examples where we use light to help us. How many can you think of?

Reflection of light

When waves meet surfaces or boundaries between substances, they can reflect off the surface and travel back through the first medium. Reflection is the process of waves bouncing off a surface and returning in the opposite direction. Water waves bounce off seawalls and sound waves reflect off surfaces as echoes. Light behaves like a wave and will reflect off certain objects, surfaces, such as when sunlight reflects off a shiny surface and is bright to your eyes. Scientists have used experiments to develop the law of reflection, which helps us to understand how reflection happens (see Figure 7.7.1).

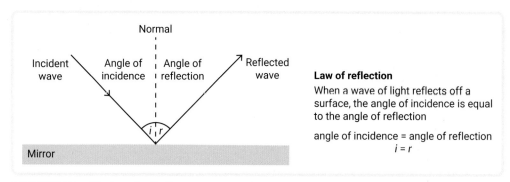

▲ **FIGURE 7.7.1**   The law of reflection

incident wave
a wave of light that is approaching a surface

reflected wave
a wave of light that has been reflected off a surface

The **incident wave** is the wave that is approaching the surface. The **reflected wave** is the wave that has been reflected off the surface. The normal is an imaginary line that is perpendicular to the surface at the point where the incident wave meets the surface.

Refraction of light

refraction
the transmission of a wave from one medium into another with a resulting change of direction

When waves travel from one medium to another, their speed and direction can be affected. **Refraction** of light is when light rays bend as they pass into a new medium. You may have noticed this with an object in water. In Figure 7.7.2, the light that is reflected off the part of the pencil in the water travels a bent path on its way to the human eye when compared to the path travelled by the light reflected off the dry part of this pencil. This is because the light coming off the bottom part of the pencil changes direction as it moves from the water into the air.

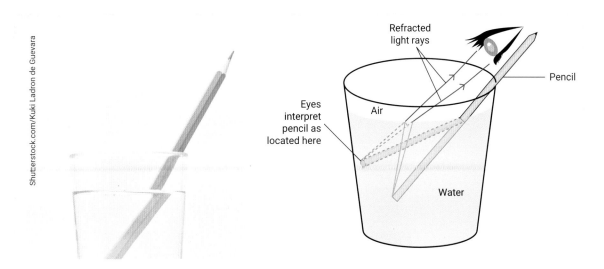

▲ **FIGURE 7.7.2**   Light undergoes refraction and bends in new directions when passing from water to air.

The refraction of light is determined by the change in optical density of the substance. When light travels from a less dense to a more dense medium, such as from air to water, the light bends towards the normal to the boundary. When light travels from a more dense to a less dense medium, the light bends away from the normal to the boundary. This phenomenon is easier to imagine by using the analogy of a car's wheels, as shown in Figure 7.7.3.

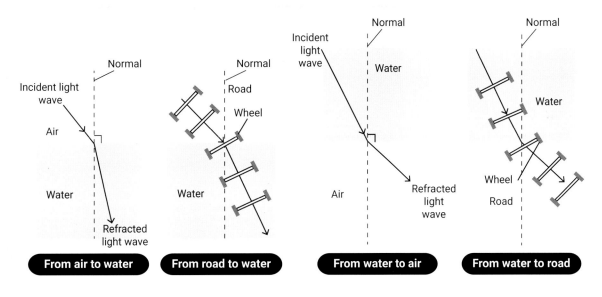

▲ **FIGURE 7.7.3**   Light changes direction when entering a new medium on an angle in the same way that a car might change direction when entering water.

When glass prisms separate white light into different colours, it is known as **dispersion**. Dispersion is based on principles of refraction. As the light travels from the air to the glass prism, different wavelengths of visible light experience different amounts of bending. This occurs again as the light leaves the prism and enters the air again, resulting in the different colours of light spreading out and travelling in separate, slightly different directions (Figure 7.7.4).

dispersion
the separation of light into its different colours

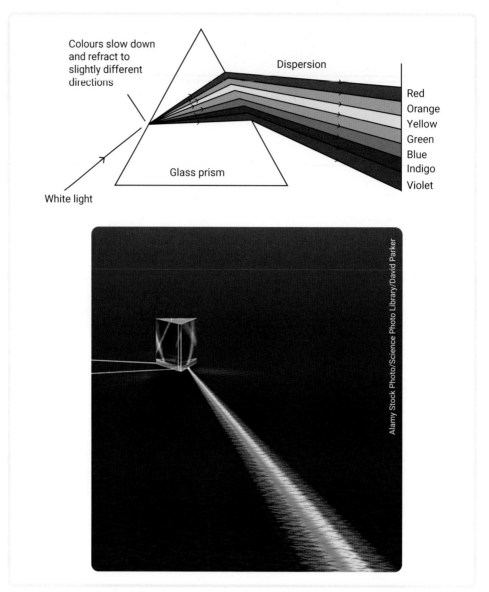

Colours slow down and refract to slightly different directions

Dispersion

Red
Orange
Yellow
Green
Blue
Indigo
Violet

Glass prism

White light

Alamy Stock Photo/Science Photo Library/David Parker

▲ FIGURE 7.7.4   The dispersion of light occurs when different wavelengths of light are bent different amounts as they are refracted when crossing boundaries between glass and air.

Light waves and seeing

We see an object because light is reflected from the object's surface and enters our eyes. The light travels through the eye and is received by cells that send signals to the brain where we interpret the information as images. Figure 7.7.5 shows how light travels through the eye.

9780170472838

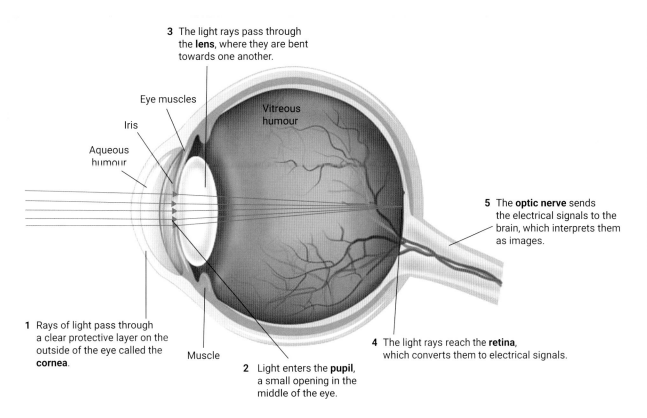

3 The light rays pass through the **lens**, where they are bent towards one another.

Eye muscles

Iris

Aqueous humour

Vitreous humour

5 The **optic nerve** sends the electrical signals to the brain, which interprets them as images.

1 Rays of light pass through a clear protective layer on the outside of the eye called the **cornea**.

Muscle

4 The light rays reach the **retina**, which converts them to electrical signals.

2 Light enters the **pupil**, a small opening in the middle of the eye.

▲ **FIGURE 7.7.5** The structure of the human eye and the pathway of light through the eye

Observing the law of reflection

☆ **ACTIVITY**

Use a single-slit light box or a laser beam to demonstrate the law of reflection. Point it on an angle at a mirror and use a protractor to measure the incident angle and reflected angle. Do not look directly at a light or laser source.

Based on your measurements, **evaluate** whether the law of reflection has been demonstrated or not.

7.7 LEARNING CHECK

1 **Draw** a diagram showing the law of reflection. On your diagram, **identify** the:
   a incident wave.
   b reflected wave.
   c normal.
   d angle of incidence.
   e angle of reflection.
2 **Define** 'refraction'.
3 **Construct** a diagram of the refraction that occurs when a wave of light travels out of water (more dense) and into air (less dense).
4 **Describe** the function of the retina in the eye.

7.8 Thermal energy and temperature

BY THE END OF THIS MODULE, YOU WILL BE ABLE TO:

✓ define and distinguish between the concepts of thermal energy, temperature and heat.

Video activity
The race for absolute zero

GET THINKING

'Heat', 'thermal energy' and 'temperature' are key words in this module. Write a sentence using each of these key terms to demonstrate your understanding before you begin.

Temperature and thermal energy

thermal energy
the energy associated with a body resulting from the kinetic energy and potential energy of particles

heat
thermal energy that is being transferred between different places or particles

The amount of **thermal energy** within matter is determined by the total kinetic energy and potential energy within the particles. When energy is added to solid particles, their movement increases. This results in an increase in kinetic energy and, hence, their thermal energy. If energy is added and the particles spread apart and the substance changes state into a liquid, the particles' thermal energy increases as a result of the potential energy increasing.

Heat is a term that is used to describe the transfer of thermal energy. It always refers to the movement of energy, rather than energy that is held somewhere or contained in certain matter. Therefore, particles have thermal energy, but it is heat that is transferred from particle to particle, or space to space. Depending on whether particles absorb or emit heat, their thermal energy will increase (they will get hotter) or decrease (they will get colder) (Figure 7.8.1).

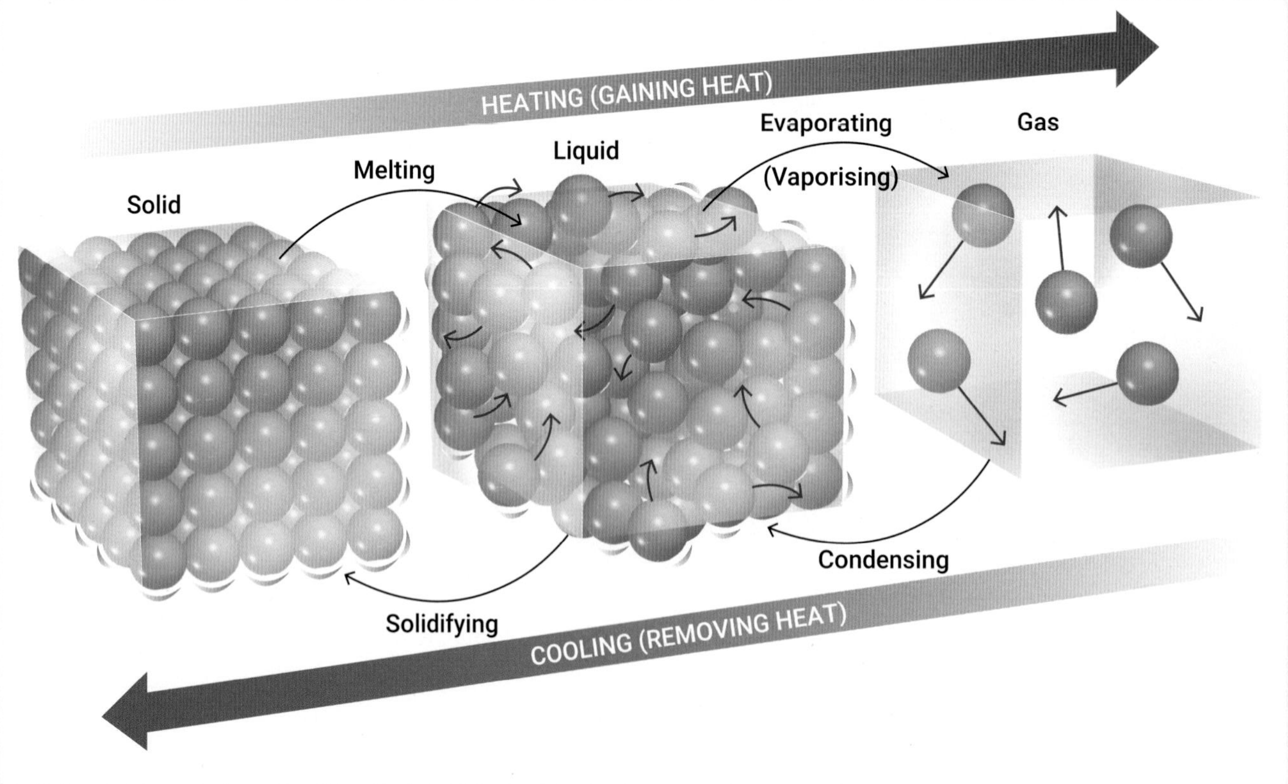

HEATING (GAINING HEAT)
Evaporating
(Vaporising)
Gas
Liquid
Melting
Solid
Condensing
Solidifying
COOLING (REMOVING HEAT)

▲ **FIGURE 7.8.1** When thermal energy is added to particles, kinetic energy and potential energy increase, leading to changes of state.

Temperature is a measurement of the kinetic energy within a substance and so can indicate the amount of thermal energy. It tells us how hot or cold something is. It is usually measured in degrees Celsius (°C). Table 7.8.1 shows some different temperatures as measured on different scales.

temperature
a measurement of the kinetic energy within a substance that lets us know how hot or cold something is

7.8

▼ TABLE 7.8.1 Some significant temperatures as measured in degrees Celsius

Temperature (°C)	Temperature (K)	Significance
−273	0	Absolute zero
0	273	Freezing point of water
25	298	Standard room temperature
37	310	Human body temperature
100	373	Boiling point of water (at sea level)

Absolute zero (−273°C) is the lowest possible temperature. At this temperature, there is no particle movement and no thermal energy (Figure 7.8.2). All substances are in solid form at this temperature. At temperatures above absolute zero, all matter has moving particles and thermal energy. 'Hot' and 'cold' are just words that describe temperature relative to something else, usually our body temperature or the average air temperature. Hot things have more thermal energy and cold things have less thermal energy.

absolute zero
the lowest temperature possible (−273°C) at which matter is considered to have no thermal energy

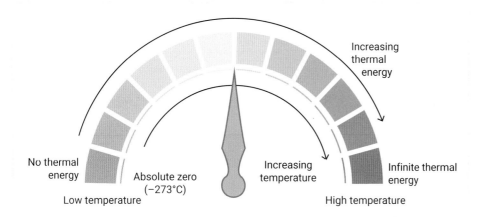

▲ FIGURE 7.8.2 Measuring temperature and thermal energy

7.8 LEARNING CHECK

1 **Distinguish** between thermal energy and heat.

2 **Explain** why cold air inside your freezer has thermal energy.

BY THE END OF THIS MODULE, YOU WILL BE ABLE TO:

✓ recall that thermal energy can be transferred by three different mechanisms and describe how each works

✓ explain heat transfer by radiation using the wave model.

Video activity
Heat transport

Interactive resource
Drag and drop:
Conduction,
convection and
radiation

Extra science investigation
Comparing thermal
conductivity

GET THINKING

Thermal energy can be transferred in many ways and examples of heating and cooling can be found everywhere. Can you make a list of 10 different ways of heating something? Can you make a list of 10 ways of cooling something?

Heat transfer

'Heat' refers to the transfer of thermal energy from one substance to another, one particle to another or one place to another. Heat transfer occurs from higher temperatures to lower temperatures. Some examples of heat transfer are given in Table 7.9.1.

▼ **TABLE 7.9.1** Some examples of heat transferring from areas of higher temperature to areas of lower temperature

Example	Description
 Holding a cup of hot liquid	• The liquid is hotter than the cup and hotter than the hands. • Heat transfers from the liquid to the cup and to the hands. • Thermal energy and the temperature of the hands increases, while the thermal energy of the liquid decreases.
 Putting an icepack on a sprained ankle	• The ankle is hotter than the icepack. • Heat is transferred from the ankle to the icepack. • As the icepack receives thermal energy, its temperature increases, and it may melt. • As the ankle loses thermal energy, its temperature decreases, and it feels cold.
 Wearing a hoodie	• A hoodie slows the heat loss to the air and traps your body heat. • The hoodie is an insulator and keeps your body temperature higher for longer, so you don't feel as cold.

There are many ways by which heat can be transferred. The three main ways are conduction, convection and radiation (Figure 7.9.1).

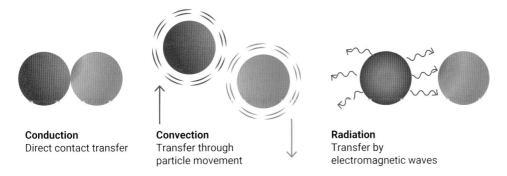

Conduction
Direct contact transfer

Convection
Transfer through particle movement

Radiation
Transfer by electromagnetic waves

▲ **FIGURE 7.9.1** Heat transfer can occur by conduction, convection or radiation.

Conduction and convection can be modelled as the transfer of energy via particles. Heat radiation is considered to be a type of electromagnetic radiation and is modelled by waves.

Conduction

Conduction is the heat transfer between particles by direct contact. Heat transfers from hotter particles to colder particles by the transfer of kinetic energy when they vibrate against one another. The faster vibrations of the hotter particles cause the colder particles to vibrate as well and then heat up. Conduction mostly occurs in solid substances because the particles are close together and can pass kinetic energy from one to another easily (Figure 7.9.2).

conduction
the transfer of heat between particles by direct contact

Some examples of conduction are the transfer of heat from:

- a frying pan to an egg to cook it
- an iron to your clothes when ironing them (Figure 7.9.3)
- hot sand to your feet when walking across it on a summer's day.

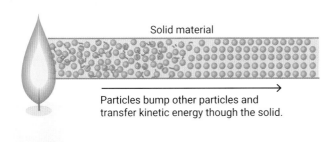

Solid material

Particles bump other particles and transfer kinetic energy though the solid.

▲ **FIGURE 7.9.2** Conduction is the process of passing thermal energy by direct contact between particles.

▲ **FIGURE 7.9.3** Heat is transferred from the iron to your clothes.

Some substances are better conductors of heat than others. Metals conduct heat quite well and, thus, easily transfer heat. Other substances such as wool and rubber are poor conductors of heat. Substances that do not transfer heat well by conduction are known as **insulators**. Any material that prevents heat from being easily transferred is an insulator (Figure 7.9.4). Foam plastic makes a good insulator for drinks. Insulating materials are used in the roofs and wall cavities of houses to reduce heat loss in winter and reduce heat gain in summer.

insulator
a material that restricts heat transfer by conduction

▲ **FIGURE 7.9.4**   Saucepans often have insulated handles that do not conduct heat, so we can pick them up, even when the rest of the pot is very hot.

Convection

convection
the transfer of heat within fluids (liquid or gas) by particle movement

density
a measure of mass per unit volume or how compact mass is within a certain space

Convection is a type of heat transfer that happens when particles move through space. Convection occurs in fluids (liquids and gases) because particles can move around. Convection cannot occur in solids because solid particles are in a fixed position and cannot move through the substance. Hotter particles and colder particles move through a fluid in different ways. Hotter particles spread out because of their increased vibration and can become less dense, whereas colder particles are more compact and denser. Because small groups of particles have different **densities**, the less dense particles (hotter particles) tend to rise up in a fluid and the denser (cooler) particles sink down, causing convection currents (Figure 7.9.5). This results in the transfer of thermal energy from lower areas to higher areas as the hotter particles travel through the fluid.

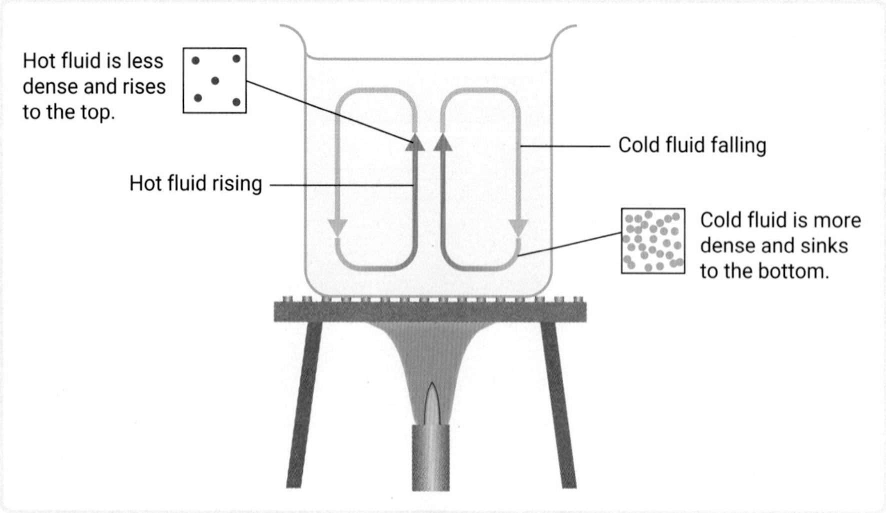

Some examples of convection heat transfer are:
- hot air rising in a room with a heater on the floor
- hot air rising in a hot air balloon and allowing the balloon to float upwards
- hot water rising to the surface in a saucepan on the stove.

Hot fluid is less dense and rises to the top.

Hot fluid rising

Cold fluid falling

Cold fluid is more dense and sinks to the bottom.

▲ **FIGURE 7.9.5**   Convection occurs when fluid particles rise and fall based on their relative densities.

9780170472838

Radiation

Infrared radiation is a type of electromagnetic radiation that has slightly lower energy than visible light. Infrared radiation can transfer thermal energy when an object absorbs infrared waves and the object's thermal energy increases. An example of radiative heat transfer is sitting by a fire and feeling the intense heat of the fire on your legs, but the air between the fire and you has not heated to the same temperature of your legs (Figure 7.9.6).

infrared radiation
electromagnetic radiation
that transfers heat
through space

▲ **FIGURE 7.9.6** Getting hot feet or legs sitting next to the fire is an example of radiative heat transfer.

All objects above –273°C give off some electromagnetic radiation, even if we can't feel it. Hotter objects emit infrared waves of a greater frequency that carry more energy than the waves emitted by colder objects. Infrared cameras can receive infrared radiation and convert different frequencies of radiation to different colours on a screen.

In Figure 7.9.7, you can see an infrared image of a heated house on a cool night. The higher temperature materials, such as the walls of the house and the windows, are emitting infrared waves with more energy than the cooler or insulated materials such as the grass, roof or night sky. The camera makes a prediction of the temperature of these materials based on the frequency of infrared radiation they are emitting.

▲ **FIGURE 7.9.7** An infrared photo of a heated home on a cool night

In the same way that visible light can be reflected, absorbed or transmitted when coming into contact with a surface, so can all types of electromagnetic waves, including infrared radiation (Figure 7.9.8).

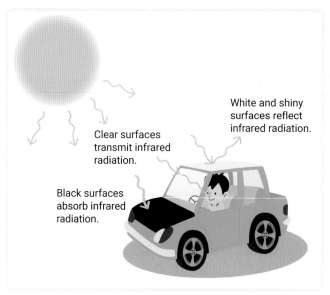

White and shiny surfaces reflect infrared radiation.

Clear surfaces transmit infrared radiation.

Black surfaces absorb infrared radiation.

▲ FIGURE 7.9.8　All electromagnetic radiation can be reflected, absorbed or transmitted by different surfaces

☆ ACTIVITY

Observing heat transfer

Place two surfaces, one that is light in colour (e.g. white) and one that is dark (e.g. black), in the sunlight for 10 minutes. Investigate whether their temperatures are different and **explain** using your understanding of heat transfer.

Shutterstock.com/ImageFlow

▲ FIGURE 7.9.9　An example of a light and a dark surface placed in the sunlight

9780170472838

1 **State** whether each statement is true or false and justify your answer.

   a Heat transfer occurs from cool particles to warmer particles.

   b Ice-cream has thermal energy.

   c Metal often feels cold because heat transfer happens quickly from our fingers to the metal.

   d In a convection current, cold air rises and warm air sinks.

   e Heat transfer by radiation requires particles.

2 **Describe** heat transfer by conduction, convection and radiation in one sentence each. Include an example of each type of heat transfer.

3 To keep houses warm, many are built with double-glazed windows that have two panes of glass with a gap of gas in between (Figure 7.9.10). **Justify** why air is a better choice than water for the insulating material between the glass, in terms of heat transfer.

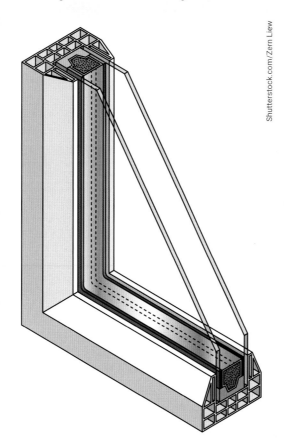

▲ **FIGURE 7.9.10** Double glazing

7.10 First Nations Australians' traditional musical, hunting and communication instruments

IN THIS MODULE, YOU WILL:

✓ investigate the materials used to transfer sound energy in First Nations Australians' instruments.

First Nations Australians' instruments

Sound is a wave that is transmitted from one location to another through a medium, such as air or water. For many thousands of years, First Nations Australians have used scientific understanding of the transfer of sound waves for hunting, communication and music.

The Quandamooka Peoples of Minjerribah (south-east Queensland) applied the knowledge of sound travelling in water to hunt fish in partnership with dolphins. Quandamooka Peoples called the dolphins by making a specific sound by slapping spears on the water surface and digging them into the sand. The sound waves alerted the dolphins, and the dolphins herded fish into waiting fishing nets. In other parts of Australia, First Nations Australians clapped stones together under water. This startles fish and brings them closer to the surface for easier capture.

▲ **FIGURE 7.10.1** (a) A bottle nose dolphin; (b) a dolphin place marker designed by artist Belinda Close in Pulan (Amity Point), Queensland, represents Quandamooka Peoples.

First Nations Australians use many instruments to generate sound waves for communication and music. The sound generated by any instrument depends on the materials used in its construction.

Idiophones are instruments that generate sound waves by vibrating themselves, without the use of strings or membranes. Instruments that are struck on each other, such as boomerangs or clapsticks, are examples of First Nations Australians idiophones. The sound produced depends on the type of wood used to construct the instrument, the length of the instrument and the position at which they are struck together.

9780170472838

Rattles and shakers are also idiophones. The sound waves are generated indirectly by the operator making two materials strike together. Shakers or rattles made from *kulaps* are produced by many of the Peoples of the Torres Strait Islands (Figure 7.10.2). *Kulaps* are the hard, brown shells of seeds of the matchbox bean vine. The shells are cut in half and attached to a length of twine to make the rattle.

Membranophones are instruments that generate sound waves through the vibration of a stretched membrane. Sound waves are produced when the stretched membrane is struck by hand or a drumstick. First Nations Australians across the continent have used skin from animals, including goanna, kangaroo, possum and lizards, to make drums. For example, the Tjungundji and Yupangathi Peoples from the Cape York Peninsula in Queensland make a drum by stretching goanna skin across a hollow pandanus stem.

Joe Sambono

▲ **FIGURE 7.10.2** A rattle made from *kulaps*, the seeds of a matchbox bean vine

Making a drum

First Nations Australians select particular materials to generate a desired sound. You will make a musical instrument and investigate how the materials used affect the sound produced.

You need

- plastic bowl
- waxed paper
- soft cloth (e.g. dishwashing cloth or damp chamois)
- stretchy fabric (e.g. Lycra)
- rubber bands, tape or zip ties

What to do

1 Cover the plastic bowl with waxed paper and fix it in place with rubber bands, tape or zip ties.
2 Tap the paper with an implement or your hand.
3 Record the sound wave produced using an app or a website such as the one at the weblink.
4 Change the material on your drum.
5 Record the sound waves again.
6 Repeat with a third material.

Weblink
Virtual oscilloscope

Making a shaker

You need

- empty plastic bottle with a lid
- uncooked rice
- uncooked popcorn
- small pebbles

What to do

1 Quarter fill the bottle with uncooked rice and seal it with the lid.
2 Shake the bottle to produce sound waves.
3 Record the sound wave produced using an app or the same weblink provided in activity 1.

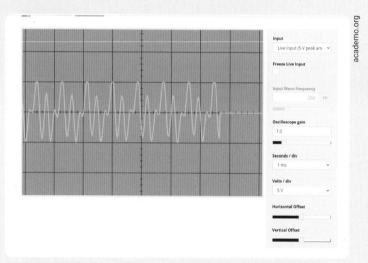

▲ **FIGURE 7.10.3** A sound wave recorded using a virtual oscilloscope.

4 Change the material inside the bottle, making sure to use the same amount.
5 Shake the bottle to produce sound waves.
6 Record the sound waves again.
7 Repeat with a third material.

What do you think?

1 How did the material used affect the sound waves produced?
2 Was this a fair test? How did you control the variables?
3 How would the sound waves be affected by the amount of material in the shaker?
4 How would the sound waves be affected by the thickness of the material used to make the drum in Activity 1?
5 How would the sound waves be affected by the tightness of the material stretched over the drum in Activity 1?

You can test these other variables to see how they affect the sound waves produced.

7.11 The development of the cochlear implant

✓ explain how a cochlear implant works to provide hearing to people with profound hearing loss.

Problems in hearing

Hearing impairment or total hearing loss affects many people. This can be caused by damage to the ear from infection, exposure to loud noises and ageing. Some people are born with partial or complete hearing loss. The first hearing aids were developed in the late 19th century. These amplified sounds as they entered the ear but didn't work for many types of hearing impairment. It wasn't until the late 1970s that scientists developed a technology that allowed sound to by-pass the ear and be transmitted directly to the auditory nerve so that the brain could hear.

Professor Graeme Clark and the development of the cochlear implant

Professor Graeme Clark is an Australian medical academic and ear, nose and throat surgeon who started researching hearing technologies in 1967. Clark faced many obstacles with his research, including lack of funds and resistance to his ideas from other scientists. Thanks to Professor Clark's persistence and hard work, he had a major success when one of his patients was able to hear speech for the first time. Working alongside doctors, government and industry professionals, Clark developed a surgical implant known as a 'cochlear implant'. This device was an affordable, effective solution that allowed hearing-impaired adults to hear. By 1990, the implant was modified and approved for use in children aged two or older.

Today, the cochlear implant is still considered to be the most significant medical advance in the history of deafness management. It is the only technology that links the sensing of stimuli directly to the human brain. More than 200 000 people over the world have benefited from the cochlear implant.

The cochlear implant bionic ear

The cochlear implant connects directly to the auditory nerve of the human ear within the cochlea. A small microphone picks up sounds outside the ear and a transmitter converts the sounds to digital signals. A receiver passes these signals through electrodes that are connected directly to the auditory nerve (Figure 7.11.1). The brain then receives the sound in the same way it would if sound was detected by the ear.

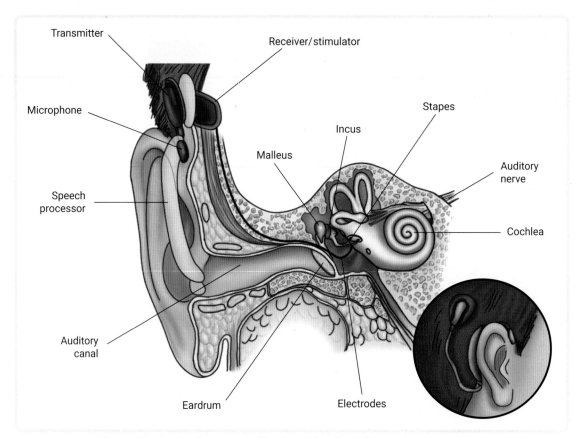

▲ FIGURE 7.11.1   The cochlear implant has a surgical implant within the cochlea and an external microphone and transmitter.

1 Before the cochlear implant was developed, people with damaged hearing used hearing aids. Research and **summarise** how:

   a a hearing aid works.

   b the hearing aid was developed.

2 **Compare** the hearing aid to the cochlear implant.

3 **Explain** why the cochlear implant would work for someone with damaged middle ear bones but a hearing aid would not.

4 **List** five main ideas from this chapter that you think would have been most important for Professor Graeme Clark and his research team to understand when developing this technology.

5 **Evaluate** how different the world may have been if the cochlear implant had never been developed.

SCIENCE SKILLS IN FOCUS

IN THIS MODULE, YOU WILL FOCUS ON
LEARNING AND IMPROVING THESE SKILLS:

▶ using a model to visualise and
understand wave behaviour

▶ evaluating the usefulness of a model by
identifying strengths and limitations.

EVALUATING SCIENTIFIC MODELS

Scientific models are not perfect. Sometimes
they do a really great job of explaining or
modelling a concept and have many strengths.
However, there are often still limitations or issues
that mean we can't use them to understand
certain ideas. Discussing these strengths and
weaknesses allows us to evaluate the quality
of a model. Evaluating scientific models lets
us determine how useful the model is and
understand how and when it can be applied.

▶ When identifying strengths of a scientific
model, considerations include:

- Which ideas or observations about the
concept being modelled does the model
show well?

- In what ways does the model explain or
help you understand observations about
the phenomena being modelled?

- Does the model predict outcomes that can
then be tested for the phenomena being
modelled?

- Is the model easy and cheap to use?

▶ When identifying limitations of a
model, considerations include:

- Are there circumstances
when the model behaves
differently from the
concept being modelled?

- What are the physical
differences between the
model and the idea?

- In what ways does the
model make it harder to
understand observations
about the phenomena
being modelled?

Video
Science skills in a
minute: Evaluating
models

**Science skills
resource**
Science skills in
practice: Evaluating
scientific models

**Extra science
investigation**
Slinky extension

There may be improvements or extensions to the
model that can be made to minimise the limitations.
However, in most cases, limitations cannot be avoided
and need to be considered when using a model. The
model may only be applicable for certain ideas relating
to the concept and not all. But it can still be very helpful
to understand these specific ideas or behaviours.

INVESTIGATION 1: MODELLING WAVES WITH SLINKYS

AIM

To model energy transfer by waves (sound and light)
using slinkys and evaluate the usefulness of the model

MATERIALS

☑ slinky that can be stretched approximately 5 m
☑ masking tape
☑ camera

METHOD

1 Work in groups of three.

2 Fix a small bit of masking tape around a central part
of the slinky wire.

3 Between two people, hold each end of the slinky
firmly and stretch it out along the ground so that it
extends 4–5 m in length.

4 One person should then move their hand from
side to side in a repeating pattern to generate a
transverse wave along the ground.

5 A third person should then take a photo or a video
of the transverse wave.

6 Make some different transverse waves by changing
the frequency of the hand movement. Make a wave
with a long wavelength (slower hand movement) and
one with a short wavelength (faster hand movement).

7 The second person on the slinky should send a
compressional or longitudinal wave along the slinky
by pulsing their hand forwards and backwards
towards the other person.

8 Take a photo or a video of the longitudinal wave.

9 Make some different longitudinal waves by
changing the frequency of the hand movement.
Make a wave with a long wavelength and one with a
short wavelength.

RESULTS

Include a copy of the pictures or videos you have taken.

ANALYSIS AND EVALUATION

1 As the transverse wave moved from one person to the other, describe the motion of the point on the slinky with the masking tape.

2 As the longitudinal wave moved from one person to the other, describe the motion of the point on the slinky with the masking tape.

3 Explain where and how the energy of each wave was generated or transferred to the slinky.

4 Considering what you know about sound waves, identify some strengths and limitations of this model for understanding sound and hearing.

5 Considering what you know about light or other electromagnetic waves, identify some strengths and limitations of this model for understanding light and seeing.

6 Suggest any improvements that could enhance the usefulness of this model.

CONCLUSION

Overall, do you think the model was a good one or not? Give reasons for your answer.

INVESTIGATION 2: USING A SIMULATION MODEL TO INVESTIGATE TRANSVERSE WAVES

AIM

To use a two-dimensional wave simulation model to investigate transverse waves

MATERIALS

☑ laptop with Internet access
☑ 'Wave on a string' simulation (MindTap)

Weblink
PhET: Wave on a string

METHOD

Setting up

1 Open the simulation and make the following adjustments to the settings in the simulation:
   • Change the wave generation mode to 'Oscillate'.
   • Change the end type to 'No end'.
   • Choose 'Slow motion'. You can change to 'Normal' at any time in this activity if preferred.
   • Turn damping to 'None'.
   • Turn on the 'Ruler' tool.

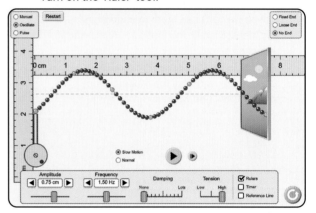

▲ **FIGURE 7.12.1** Screenshot of simulation set-up

2 Press the pause button and practise using the ruler tool to measure the wavelength of one full wave. You can click and drag the ruler tool to where you need it.

3 Un-pause the simulation so that it is running and ready for Part A.

Part A: Relationship between amplitude and wavelength

How does a change in amplitude affect the wavelength of a transverse wave when modelled on a string?

4 In your workbook or on your device, set up a results table for an investigation into how amplitude (independent variable) affects the wavelength of the wave (dependent variable). An example table has been provided in Table 7.12.1.

5 Complete the first row of the table for the starting amplitude and wavelength of the wave.

6 Increase the amplitude of the wave by 0.10 cm to 0.85 cm.

7 Measure and record the wavelength of the wave in the results table. Do this as you have done in the 'Setting up' stage.

8 Repeat steps 6 and 7 four more times to a maximum amplitude of 1.25 cm. Don't forget to unpause the simulation before making each change.

Part B: A second relationship

9 Choose another variable (frequency or tension) to investigate how it might affect the wavelength.

10 Write a research question similar to that in Part A to reflect the direction of your investigation.

11 Set up a results table for this second research question.

12 Complete the first row of the table for the starting settings of your simulation.

13 Make four changes to your independent variable and record the resulting wavelengths.

14 It is best to sequence the rows of your results table so that the independent variable is increasing. This will make it easier to identify trends.

RESULTS

1 Complete the results table for Part A (Table 7.12.1).

▼ TABLE 7.12.1 Part A results table: amplitude and wavelength data for a transverse wave

Amplitude (cm)	Wavelength (cm)

2 Include your results table for Part B here.

ANALYSIS AND EVALUATION

1 Complete the following sentence to identify a relationship in Part A.

As the amplitude of the wave increased, the wavelength _____.

2 Identify the relationship for the data collected in Part B using a similar sentence as Question 1.

3 Discuss how this simulation has been helpful to understand relationships between variables for transverse waves.

4 Suggest how this simulation might need to change if it were to model longitudinal waves instead.

CONCLUSION

Summarise the three main things you have learned about waves and simulations after completing this inquiry activity.

INVESTIGATION 3: SPEED OF SOUND WAVES

AIM

To use echoes to calculate the speed of sound

MATERIALS

☑ stopwatch
☑ large reflecting surface, preferably outdoors
☑ 2 wooden blocks
☑ tape measure or trundle wheel

METHOD

1 Work in pairs. Stand at a measured distance, s, of 50–100 m or more from the wall.

2 Clap the wooden blocks together to make a loud sound – the louder the better! Listen for the echo.

3 Next, clap the wooden blocks regularly and at quite a fast rate. Adjust the clapping rate until the echo of each clap overlaps with the next clap.

4 Use the stopwatch to measure the time taken for 11 claps at this rate.

RESULTS

1 Record the distance to the reflecting surface.

2 Calculate the distance to the reflecting surface and back to the clapper (2*s*).

3 Record the time taken for 11 claps at the clapping rate that has the claps coinciding with the echoes.

ANALYSIS AND EVALUATION

1 Calculate the time taken for one clap to travel to the wall and back (*t*). Do this by dividing the time recorded in step 3 of the Results section by 10. (We divide by 10 because it takes 10-time intervals for 11 claps to coincide with the echoes.)

2 Calculate the speed of sound by dividing the distance travelled by the sound by the time taken to cover that distance.

3 Calculate the speed of sound, *c*:

$$\text{speed of sound} = \frac{2s}{t}$$

CONCLUSION

1 Provide a realistic estimate of the speed of sound.

2 Compare your value for the speed of sound in air to the average value of 340 m/s.

EXTENSION

The speed of sound is different in different media (Table 7.12.2).

▼ TABLE 7.12.2   The speed of sound in different media

Material	Speed of sound (m/s)
Rubber	60
Air at 20°C	343
Air at 40°C	355
Water	1481
Gold	3240
Glass	4540
Copper	4600
Aluminium	6320

1 Why do you think the speed of sound is different in different media?

2 Give one reason why sound travels faster in warmer air.

3 Explain why sound travels faster through:

  a water than air.

  b gold than water.

REMEMBERING

1 **Copy and label** the following transverse wave diagram.

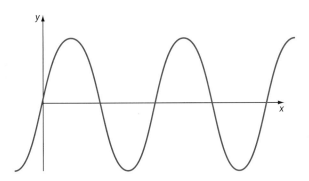

2 **Draw** a diagram of a longitudinal wave to model the action of the wave.

3 **Describe** the transfer of sound energy through a medium.

4 **Copy and annotate** the following diagram of the human eye to show how light travels through the eye.

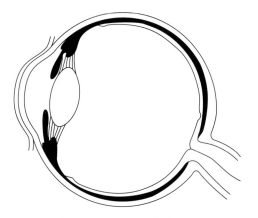

UNDERSTANDING

5 **Describe** the purpose of a scientific model, using an example.

6 **Describe** what happens to sound energy when it meets the ear drum in the human ear.

7 **Explain** how a microwave oven heats food.

8 **Describe** the pathway of light waves through the human eye by listing, in order, the structures that they pass through.

APPLYING

9 A longitudinal wave has a compression zone every 0.005 m and 20 compression zones pass by a particular point every second. **Calculate** the speed of the longitudinal wave.

10 **Describe** the changes to the speed of sound as a sound wave travels from the air to a brick wall. Give reasons for your ideas.

11 During an accident, the malleus inside your ear is broken and no longer functions. **Predict** the possible consequences of this.

12 **Complete** this sentence: When light travels from air to water, it bends _____ the normal.

ANALYSING

13 Use the wave graph to answer the following questions.

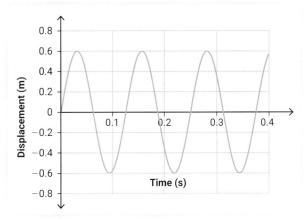

a **Identify** the period of the wave.

b **Calculate** the frequency of the wave.

c **Explain** why the wavelength cannot be determined for this wave.

14 Infrared cameras are used by emergency rescue workers looking for lost bushwalkers or skiers. **Explain** why these cameras work best in winter and at night when looking for lost people.

15 **Explain** how the Quandamooka Peoples' traditional method of hunting fish relates to their knowledge of sound.

16 a Describe the flow of heat energy if you were to grab a hot metal handle of a pot on the stovetop.

b Describe the flow of heat energy as you treat the burned area by running cold water over it.

17 If you wanted to model a low-volume high-pitched sound wave using a slinky, what would you have to do to the slinky? **Justify** your reasoning.

EVALUATING

18 Cataracts occur when the eye lens becomes cloudy or opaque and this can be caused by excessive UV radiation. **Discuss** whether fashion sunglasses without UV protection should be sold in Australia.

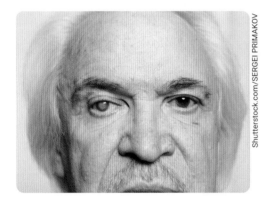

Shutterstock.com/SERGEI PRIMAKOV

19 Evaluate whether it would be safer to stir boiling stew with a wooden spoon or a metal spoon. Give reasons for your choice.

20 Scientists use both the wave model and the particle model to explain what light energy is and how it moves. Using this example, **explain** the benefits and limitations of using models to explain scientific phenomena.

CREATING

21 Use materials from around the house to make a model of either conduction, convection or radiation that could be used to explain how it works to transfer heat. **Write** an explanation for how the objects model the heat transfer and then evaluate your model.

22 Create a generalised flow chart for the transfer of energy through eyes and ears. **Annotate** the flow chart with specific examples within the eye and the ear.

23 Draw a picture of a cup of tea sitting on a bench beside an open refrigerator. Imagine your drawing is an infrared photo where the infrared radiation from each object has been converted to colours of the rainbow where the blue represents low-energy radiation and red is high-energy radiation, as shown in the following colour spectrum.

Low-energy radiation   High-energy radiation

Colour in your drawing as it would appear in an infrared photo. You can add other household or kitchen items if you wish. Some suggestions are:

- a freshly boiled kettle
- a cat lying on the floor
- bread sitting on the bench
- snow falling outside the window.

BIG SCIENCE CHALLENGE PROJECT #7

1 Connect what you've learned

In this chapter, you have learned about scientific models for energy transfer, waves, sound, light and heat. Create a mind map to show how the main ideas you have learned about are connected.

2 Check your thinking

Consider the photo of acoustic highway barriers at the beginning of the chapter and the image of the person on this page. Considering what you now know about sound, hearing and the behaviour of waves, compare the two noise protection methods and explain how they are similar and how they are different.

3 Make an action plan

For each of the sounds you identified at the start of the chapter as having negative impacts on you, other people or animals:

- explain why the sound is harmful in this context

- suggest one way in which the negative impact could be reduced. It might be a strategy to reduce the noise produced or the effect of the noise

- describe the benefits of the strategy you are suggesting.

4 Communicate

Consider one environment where noise pollution might be a problem. Create a flyer that could be posted in this environment to encourage people to reduce the amount of noise created in this place. Include information about why the noise is a problem and how it could be reduced.

Shutterstock.com/doomu

8 Conservation of energy

8.1 **Law of conservation of energy** (p. 266)

Energy cannot be created or destroyed, only transferred and transformed.

8.2 **Visualising energy transformations** (p. 270)

A Sankey diagram is a type of flow chart that uses different-sized arrows to show the amount of energy transferring or transforming through a system.

8.3 **Calculating energy efficiency** (p. 274)

Some energy transformations are more efficient than others and this can be measured.

8.4 **Thermal waste energy** (p. 278)

During energy transformations, some energy is not transformed into useful energy and is considered to be wasted.

8.5 **Reducing waste energy** (p. 282)

The energy efficiency of processes can be increased by reducing or reusing waste energy.

8.6 FIRST NATIONS SCIENCE CONTEXTS: **Heat transfer and conservation in First Nations Australians' clothing and bedding** (p. 284)

Many First Nations Australians have long made clothes and bedding to control heat loss.

8.7 SCIENCE AS HUMAN ENDEAVOUR: **Solar energy in Australia** (p. 286)

In Australia, there are different reasons individuals, industries and communities adopt solar panels.

8.8 SCIENCE INVESTIGATIONS: **Evaluating secondary evidence** (p. 288)

1 Evaluating evidence to choose an energy-efficient lighting solution

2 Measuring the efficiency of a ball bounce

9780170472838

▲ **FIGURE 8.0.1** Adults and school students protesting for climate action

In 2018 and 2019, school students all over the world participated in strikes to demand action from politicians in the fight against climate change. A major discussion point in the ongoing climate change debate is how nations source their energy, and how sustainable these methods of electricity production are. Climate advocates are fighting for the decarbonisation of energy. This would mean a complete shift from carbon-emitting energy processes such as coal-fired electricity to renewable sources such as solar and wind power.

▶ When you look around you, where can you see energy being produced and where can you see it being used?

▶ Why do you think sourcing energy is such a controversial topic in Australia?

#8 SCIENCE CHALLENGE ACCEPTED!

At the end of this chapter, you can complete the Big Science Challenge Project #8. You can use the information you learn in this chapter to complete the project.

Assessments
- Prior knowledge quiz
- Chapter review questions
- End-of-chapter test
- Portfolio assessment task: Data test

Videos
- Science skills in a minute: Secondary evidence (8.8)
- Video activities: Forms of energy (8.1); Energy transformation (8.1) ; Energy efficiency ratings (8.3); Improving energy efficiency at home (8.5); Improving solar energy technology (8.7); Future of solar energy (8.7)

Science skills resources
- Science skills in practice: Evaluating secondary evidence (8.8)
- Extra science investigation: Energy efficiency (8.3)

Interactive resources
- Label: Sankey diagrams (8.2)
- Match: Calculating energy efficiency (8.3)
- Simulation: Energy skate park (8.4)

✹ **Nelson MindTap**

To access these resources and many more, visit:
cengage.com.au/nelsonmindtap

BY THE END OF THIS MODULE, YOU WILL BE ABLE TO:

✓ recall and explain the law of conservation of energy and understand how it applies to energy transfers and transformations

✓ model energy transfers and transformations using energy flow diagrams.

Video activities
Forms of energy
Energy transformation

GET THINKING

Make a list of the possible *sources* of energy in a school science laboratory. Make a second list of all the different ways energy is *used* in a school science laboratory.

Types of energy

Energy is important in everything that we do. It exists in many different forms, some of which are shown in Table 8.1.1. The general definition for energy is the ability to do work. It can be thought of as the ability or capacity to make something happen. We measure energy in joules (J), kilojoules (kJ) or megajoules (MJ).

▼ **TABLE 8.1.1** Some different forms of energy

Form of energy	Definition	Example of performing work
Kinetic energy	The energy of something that has mass and is moving	Air particles in wind contain kinetic energy because they are moving and can turn turbines.
Thermal energy	The energy associated with the temperature and kinetic energy of particles	A hot stovetop full of thermal energy may produce steam that moves away from the stove.
Sound energy	The energy associated with the vibrations of matter caused by sound waves	A sound makes your eardrums vibrate.
Light and radiation energy	The energy carried by electromagnetic waves	Light from the Sun is converted into electrical energy in a solar panel.
Electrical energy	The energy carried by moving charges (electrons)	Electrical energy flowing to a TV can produce light through the screen.
Gravitational potential energy	The energy stored within something as a result of its position when under the influence of a gravitational force	Objects that are dropped from heights cause damage.
Elastic potential energy	The energy stored within something when it is stretched or compressed away from its natural resting position	A stretched rubber band flings an object across a room when released.
Chemical potential energy	The energy stored within a chemical, which can be released when the bonds between atoms change during a chemical reaction	When wood is burned, chemical energy is released as heat and light.
Nuclear potential energy	The energy stored within an atom's nucleus, which can be released during a nuclear reaction	At nuclear power plants, nuclear reactions release energy from radioactive nuclei.

Law of conservation of energy

When modelling and understanding where energy comes from and how it moves around, an important fundamental law of physics is the **law of conservation of energy**. This law states that *energy cannot be created or destroyed*. This means that energy can only be transferred or transformed. Therefore, when there is an:

law of conservation of energy
when energy is transferred or transformed, the total amount of energy remains the same

- energy increase, the energy must have come from somewhere or something else
- energy decrease, the energy must have gone to somewhere or something else.

Imagine a person diving from a tall platform into a pool (Figure 8.1.1). At the top of the platform, the diver has gravitational potential energy but no kinetic energy because they are not moving. At the bottom of the dive when the diver meets the water, the diver has less gravitational potential energy, but they have a lot of kinetic energy because they are moving very fast. The lost gravitational potential energy has converted into kinetic energy as the diver falls.

According to the law of conservation of energy, the total amount of energy that exists always stays the same.

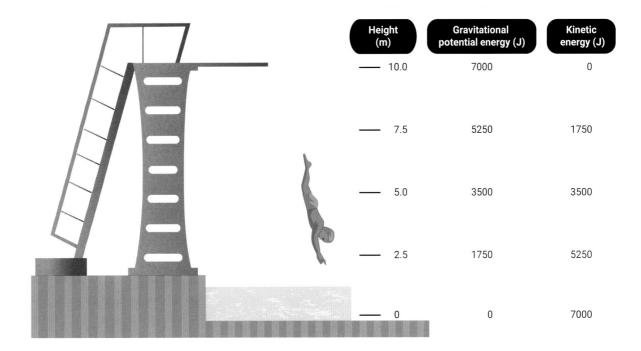

Height (m)	Gravitational potential energy (J)	Kinetic energy (J)
10.0	7000	0
7.5	5250	1750
5.0	3500	3500
2.5	1750	5250
0	0	7000

▲ FIGURE 8.1.1   A diver loses gravitational potential energy and gains kinetic energy.

Energy can be moved around in two ways: energy transfer and energy transformation.

Energy transfers

An energy transfer is when a particular form of energy is passed from particle to particle, object to object or space to space without the type of energy changing. For example:

- kinetic energy in a cricket bat transferring to kinetic energy in a cricket ball when the ball is hit
- thermal energy transferring from a hot iron to a shirt when ironing
- sound energy transferring from a speaker through air particles to your ears.

In Chapter 7, you learned about the way in which energy is transferred, with a focus on sound, light and thermal energy.

▲ FIGURE 8.1.2   A moving cricket bat transfers kinetic energy to a cricket ball when it hits it.

Energy transformations

An energy transformation happens when energy changes from one form into another. An example of an energy transformation is when a diver's gravitational potential energy changes to kinetic energy as they fall into a pool. Some other examples of energy transformations are:

- electrical energy turning into light and thermal energy in a light bulb
- light energy changing into electrical energy in a solar panel (Figure 8.1.3).
- chemical potential energy from food transforming to thermal energy.

◄ FIGURE 8.1.3   A solar panel transforms light energy into electrical energy.

Energy flow diagrams

energy flow diagram
a visual representation of energy movement using arrows and boxes

An **energy flow diagram** shows energy transfers and transformations. Energy flow diagrams use arrows to show the movement of energy, with transfers and transformations often distinguished by different colours. An energy flow diagram for a parent pushing a child on a swing is shown in Figure 8.1.4.

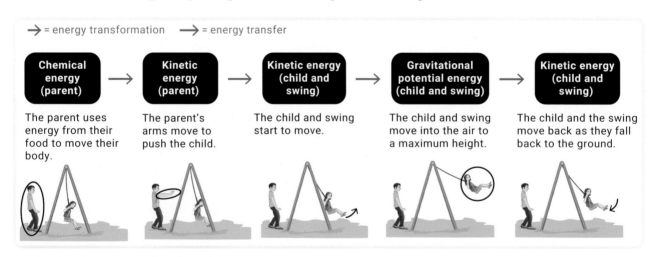

▲ FIGURE 8.1.4   An energy flow diagram for a parent pushing a child on a swing

9780170472838

Figure 8.1.5 shows a simple energy flow diagram for the transformations of energy in the production of electricity in a hydroelectric power plant.

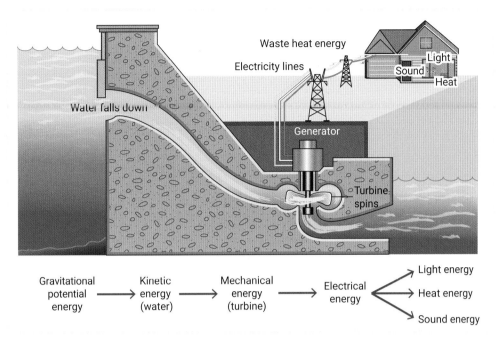

Gravitational potential energy $\longrightarrow$ Kinetic energy (water) $\longrightarrow$ Mechanical energy (turbine) $\longrightarrow$ Electrical energy $\longrightarrow$ Light energy / Heat energy / Sound energy

▲ FIGURE 8.1.5   In a hydroelectric power plant, the gravitational potential energy of water stored at height is transformed into kinetic energy and then electrical energy.

8.1 LEARNING CHECK

1 **Distinguish** between an energy transfer and an energy transformation.

2 Copy and complete the following table. For each example, **identify** the type of energy input and output and then **classify** whether this is an example of an energy transfer or energy transformation. The first example has been done for you.

Situation	Energy input type	Energy output type	Transfer or transformation?
Burning wood in a campfire	Chemical potential energy (in wood)	Thermal and light energy (flames)	Transformation
Riding a seesaw			
Watching television			
Hitting a golf ball			
Doing a bomb dive in a pool			
Heating a pie in an oven			
Sling-shotting a rock			
Boiling water on a stove			

3 **Construct** an energy flow diagram for using an electric stove to fry an egg.

4 As a car brakes to a stop, it starts with a maximum amount of kinetic energy and ends with none. Considering the law of conservation of energy, **explain** how this happens and where the energy goes.

BY THE END OF THIS MODULE, YOU WILL BE ABLE TO:

✓ construct and interpret Sankey diagrams to model different energy transfers and transformations.

Interactive resource
Label: Sankey diagrams

GET THINKING

This module shows you how to use diagrams to visualise and understand energy transformations. What are some other examples of where diagrams can be helpful for understanding?

Useful and waste energy

We use energy transfers and transformations to make things happen. When an energy transfer or transformation is intentional, it produces the type of energy you need or want. For example, when you throw a ball, you want the ball to have kinetic energy when you release it. This energy is referred to as **useful output energy**. The useful output energy for a light bulb is light and the useful output energy for a radiator is heat. In all energy transfers and transformations, there are also some unwanted energy transfers and transformations that cannot be avoided. The energy produced by these is known as **waste output energy**. This waste energy is often heat, such as a light bulb getting warm, or sound, such as a hairdryer producing noise (Figure 8.2.1).

▲ **FIGURE 8.2.1** A hairdryer produces waste sound energy when it blows hot air.

useful output energy
the output energy of a process or action that is intended and useful

waste output energy
the output energy of a process or an action that is not intended and not useful for the main purpose of the process or action

Some other examples of useful and waste energy are given in Table 8.2.1.

▼ **TABLE 8.2.1** Some examples of useful and waste energy in different situations

Process	Input energy	Useful output energy	Waste output energy
Throwing a ball	Chemical potential energy	Kinetic energy	Thermal energy of air; sound of ball moving
Using a light bulb	Electrical energy	Light energy	Thermal energy of warm bulb
Boiling water in a kettle	Electrical energy	Thermal energy of water	Thermal energy of steam
Watching television	Electrical energy	Light and sound energy	Thermal energy of television
Wind turbine	Kinetic energy of wind	Kinetic energy of turbine	Thermal energy of air, sound of moving blades

Sankey diagrams

A **Sankey diagram** is an energy flow diagram that:

- shows measurements of more than one output energy during an energy transfer or transformation
- identifies each output energy as useful or waste
- quantifies energy amounts and shows proportions of energy that flow in different directions

Sankey diagrams show flows of energy as arrows, where the thickness of the arrow indicates how much energy there is. When energy is transferred or transformed, the arrow splits into multiple arrows. Useful energy transfers and transformations are shown as arrows going left to right. Waste energy flows are directed downwards. Different energy types are sometimes shown in different colours.

To satisfy the law of conservation of energy, the total thickness of output arrows must always add up to the total thickness of the input arrows.

Figure 8.2.2 shows a Sankey diagram for a light bulb. The input energy is shown as 200 J of electrical energy. Because light is the useful output energy, the transformation with light as an output is represented horizontally. Thermal energy is a waste energy output of this energy transformation and so is drawn in the downwards direction. The useful light energy is 150 J, which is three-quarters or 75 per cent of the input energy. Therefore, the light energy arrow is three-quarters the size of the input electrical energy arrow. The waste thermal output energy is 50 J, which is one-quarter or 25 per cent of the input energy. Therefore, the downwards arrow is one-quarter of the size of the input energy arrow.

Sankey diagram
a type of flow chart that uses arrows of various sizes to indicate the amount of energy transferring or transforming in a system

8.2

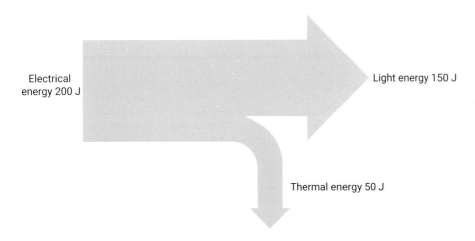

▲ FIGURE 8.2.2   A Sankey diagram for a 75 per cent efficient light bulb that uses 200 J of electrical energy

If you think about a bouncing ball, you'll notice the height of the ball's bounce decreases with each bounce, as shown in Figure 8.2.3. This occurs as more energy is transferred and transformed and waste energy is produced.

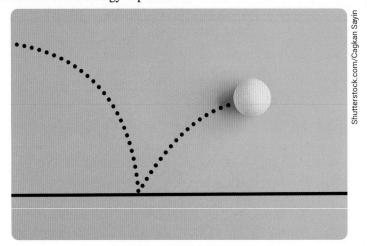

▲ **FIGURE 8.2.3** The height of a ball's bounce becomes lower with each bounce as energy is transferred and transformed.

A Sankey diagram for a ball bouncing is shown in Figure 8.2.4. This diagram shows multiple energy transfers and transformations in the process rather than a single one. The different types of energy (potential, kinetic, elastic and thermal energy) are shown in different colours. The waste energy is again represented by the downwards arrows.

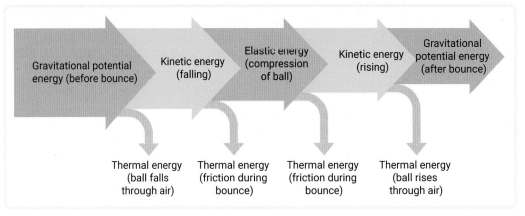

▲ **FIGURE 8.2.4** A Sankey diagram for a ball bouncing

You'll see that the thickness of the arrows decreases from left to right. The decrease in the ball's energy, which we observe in the decreased height of each bounce, is visually represented by the reduction in the width of the arrows.

9780170472838

1 **Identify** the useful output energy and a possible waste output energy for a:

a washing machine.

b toaster.

c blender.

▲ FIGURE 8.2.5   Identify the useful output energy and waste output energy of these devices.

2 **Construct** a Sankey diagram to show the energy transformation for a kettle that uses 80 kJ and produces 60 kJ of thermal energy.

3 A hairdryer converts 600 J of energy into 300 J of kinetic energy, 200 J of thermal energy and 100 J of sound energy.

a **Identify** which type of energy is the wasted energy.

b **Construct** a Sankey diagram to represent the energy transformations.

4 For the Sankey diagram in Figure 8.2.6, **identify** the:

a amount of waste energy.

b amount of useful output energy.

c efficiency of the process.

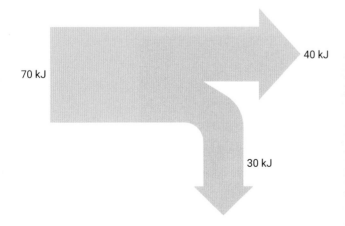

70 kJ

40 kJ

30 kJ

▲ FIGURE 8.2.6   A Sankey diagram

✓ define the concept of energy efficiency and calculate energy efficiency for different processes

✓ compare the energy efficiencies of appliances and devices and describe how a star rating can be used to indicate energy efficiency.

GET THINKING

What does the term 'efficiency' mean to you? How do you think it might be relevant to ideas about energy and energy transformations? Write down some ideas before you start this module.

Energy efficiency

energy efficiency
a measure of how much input energy is converted to useful output energy, often stated as a percentage

Energy efficiency is a measure of how much input energy is converted into useful energy. Processes that are highly efficient convert a large amount of input energy into useful output energy and have small amounts of waste energy. Low-efficiency devices or processes produce a lot of waste energy.

Energy efficiency is usually measured as a percentage and is found by calculating the percentage of the input energy that is converted into useful output energy:

$$\text{Efficiency} = \frac{\text{useful output energy}}{\text{total input energy}} \times 100\%$$

Video activity
Energy efficiency ratings

Interactive resource
Match: Calculating energy efficiency

Extra science investigation
Energy efficiency

Calculating energy efficiency

The following worked examples show how to calculate the energy efficiency of a light bulb (Worked example 8.3.1) and an electric stove stop (Worked example 8.3.2).

WORKED EXAMPLE 8.3.1

Calculate the energy efficiency of a light bulb that has a total input energy of 200 J and a useful output energy of 150 J.

◀ **FIGURE 8.3.1** A light bulb converts electrical energy to light energy.

THINKING PROCESS	WORKING
Step 1. Identify the known and unknown variables.	Total input energy = 200 J Useful output energy = 150 J Efficiency = ?
Step 2. Identify the appropriate relationship.	$\text{Efficiency} = \dfrac{\text{useful output energy}}{\text{total input energy}} \times 100\%$
Step 3. Substitute known values.	$\text{Efficiency} = \dfrac{150}{200} \times 100\%$
Step 4. Rearrange, using algebra if necessary, and solve.	Efficiency = 75% The light bulb is 75% energy efficient.

WORKED EXAMPLE 8.3.2

When preparing dinner, Charlie heats up a frying pan on an electric stove, which uses 800 J of electrical energy. Calculate the efficiency of heating the pan on the stove if:

- 400 J of thermal energy is absorbed by the pan

- 100 J of sound energy is produced

- 300 J of thermal energy is lost to the air and other parts of the stove top.

◀ FIGURE 8.3.2   Electric stoves convert electrical energy to thermal energy to cook food.

THINKING PROCESS	WORKING
Step 1. Identify the known and unknown variables.	Total input energy = 800 J Useful output energy = 400 J Efficiency = ?
Step 2. Identify the appropriate relationship.	$\text{Efficiency} = \dfrac{\text{useful output energy}}{\text{total input energy}} \times 100\%$
Step 3. Substitute known values.	$\text{Efficiency} = \dfrac{400}{800} \times 100\%$
Step 4. Rearrange, using algebra if necessary, and solve.	Efficiency = 50% The stove to pan heating process is only 50% energy efficient.

Calculating useful output and waste energy

If the energy efficiency of a device or process is given with the total input energy, you can calculate the useful output energy and waste energy from the following equations:

$$\text{Useful output energy} = \frac{\text{efficiency} \times \text{total input energy}}{100}$$

$$\text{Total input energy} = \text{useful output energy} + \text{waste energy}$$

WORKED EXAMPLE 8.3.3

A kettle in the school staff room uses 70 kJ of electrical energy and is only 80 per cent efficient. Calculate how much waste energy is produced by the kettle.

To calculate this, we need to break this example down into two parts.

PART A: CALCULATE THE USEFUL OUTPUT ENERGY

THINKING PROCESS	WORKING
Step 1. Identify the known and unknown variables.	Total input energy = 70 kJ = 70 kJ $\times \dfrac{1000\text{ J}}{1\text{ kJ}}$ = 70 000 J Efficiency = 80% Useful output energy = ?
Step 2. Identify the appropriate relationship.	Useful output energy = $\dfrac{\text{efficiency} \times \text{total input energy}}{100}$
Step 3. Substitute known values.	Useful output energy = $\dfrac{80 \times 70\,000}{100}$
Step 4. Rearrange, using algebra if necessary, and solve.	Useful output energy = 56 000 J

PART B: CALCULATE THE WASTE OUTPUT ENERGY

THINKING PROCESS	WORKING
Step 1. Identify the known and unknown variables.	Useful output energy = 56 000 J Total input energy = 70 000 J Waste output energy = ?
Step 2. Identify the appropriate relationship.	Total input energy = useful output energy + waste output energy
Step 3. Substitute known values.	70 000 = 56 000 + waste output energy
Step 4. Rearrange, using algebra if necessary, and solve.	Waste output energy = 70 000 − 56 000 Waste output energy = 14 000 J $\times \dfrac{1\text{ kJ}}{1000\text{ J}}$ = 14 kJ There is 14 kJ of waste energy.

9780170472838

Comparing efficiency

Incandescent light bulbs used to be the most common types of light bulbs but have been mostly phased out because of their low energy efficiency. Compact fluorescent lights (CFLs) and light-emitting diodes (LEDs) are now common. They have higher efficiency, which means they use less electricity to produce the same amount of light (Figure 8.3.3). People choose more efficient light bulbs to save electricity, to save money and for environmental reasons such as reducing carbon emissions.

▲ **FIGURE 8.3.3** Different types of light bulbs have different energy efficiencies. **(a)** An incandescent light bulb is 10–20 per cent energy efficient. **(b)** A compact fluorescent light bulb is 70–85 per cent energy efficient. **(c)** A light-emitting diode is 80–90 per cent energy efficient.

The Australian Energy Rating Label scheme allows people to compare the energy efficiencies of different appliances. The star rating shows the efficiency of the appliance compared with other appliances of a similar size. The more stars, the more efficient the appliance. Six stars means that the appliance is one of the most efficient options available (Figure 8.3.4).

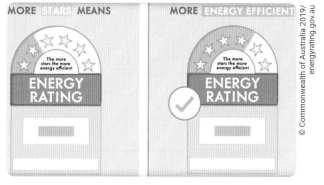

▲ **FIGURE 8.3.4** The energy rating label has a star rating system to show the efficiency of different appliances.

8.3 LEARNING CHECK

1 **Calculate** the efficiency of a torch that converts 100 J of electrical energy and produces 40 J of waste thermal energy.

2 During an all-day movie marathon, a remote control uses 1500 J of electrical energy from the batteries and is only 40 per cent efficient. **Calculate** the:

   a useful output energy.

   b waste output energy.

3 Find the energy rating labels on the appliances in your house, such as the washing machine, dishwater, fridge and dryer. **Compare** the star ratings and average energy consumption and **evaluate** which appliances are the most efficient.

GET THINKING

Think about one meal that you know how to make. From start to finish, identify all the waste materials or energy that might be produced during the process of making your meal. Are there any ways you could avoid producing this waste?

Thermal waste energy

frictional heat loss
waste thermal energy produced as a result of the vibration of particles when moving past each other in a process where thermal energy production is not intended or useful

friction
a resistance force that results when two surfaces rub against one another

Interactive resources
Simulation: Energy skate park

When a ball rolls to a stop along the ground, it can seem as though the kinetic energy is leaving the ball and disappearing. However, the law of conservation of energy says that this is not the case. During many energy transfers or transformations, thermal energy can be produced as waste output energy. This is due to the interactions between particles. As the ball rolls, the interaction between the ball's surface and the ground makes particles vibrate, which increases their temperature and produces thermal energy (Figure 8.4.1). This production of thermal energy is often referred to as **frictional heat loss** because the force of **friction** is being overcome when particles rub past each other.

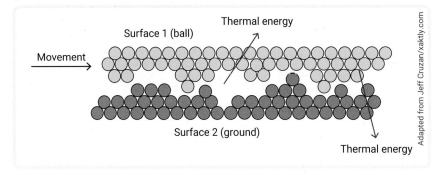

Adapted from Jeff Cruzan/xaktly.com

▲ **FIGURE 8.4.1** Thermal energy is produced when particles on the ball's surface and the ground rub against each other and vibrate.

Another example of this is when you rub your hands together. The kinetic energy from the movement of your hands is transformed to thermal energy as the particles in your skin vibrate when rubbed together. Consider the diver shown in Figure 8.4.2. The energy transformation from gravitational potential energy to kinetic energy includes some thermal energy as the air particles vibrate when the diver moves through the air. Thermal energy is produced as the diver and air particles vibrate against one another. We often hear this described as 'air resistance'. This means that the diver isn't going quite as fast when they hit the water as they might have been if there were no frictional heat losses along the way.

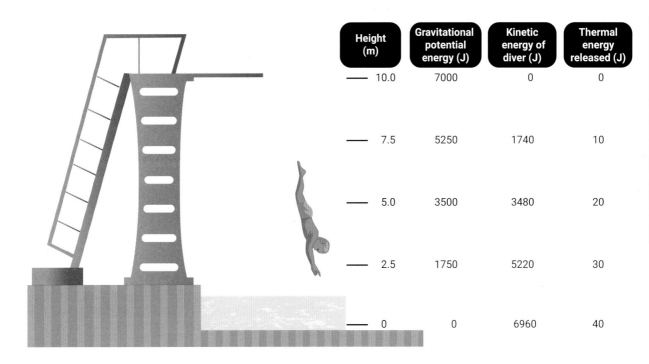

Height (m)	Gravitational potential energy (J)	Kinetic energy of diver (J)	Thermal energy released (J)
10.0	7000	0	0
7.5	5250	1740	10
5.0	3500	3480	20
2.5	1750	5220	30
0	0	6960	40

▲ **FIGURE 8.4.2** A diver loses gravitational potential energy and gains kinetic energy while producing some thermal energy as waste.

When meteors or other objects enter the atmosphere, they travel so fast through the air particles that the frictional heat losses produced can be so great that they burst into flames. Often, they burn up completely before they hit the ground. We see this happening as shooting stars in the night sky (Figure 8.4.3).

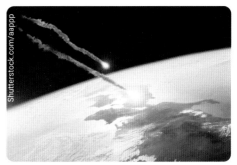

▲ **FIGURE 8.4.3** Fast-moving meteors burn up in Earth's atmosphere and appear as shooting stars.

Many electrical devices or appliances produce thermal energy that we don't need or use when transforming electrical energy into other forms such as light, sound and kinetic energy (Figure 8.4.4). This is why light bulbs, speakers, chargers, televisions, mobile phones and other devices often get warm.

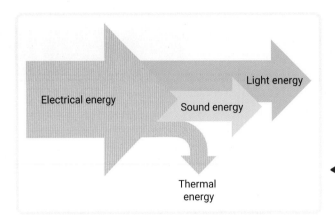

◀ FIGURE 8.4.4   A television converts electrical energy to sound and light energy with some waste thermal energy produced.

When the purpose of an energy transformation is to produce thermal energy, sometimes thermal energy can still be considered a waste output energy when it is transferred to other objects or particles. For example, boiling water on the stove involves losing thermal energy to the stovetop, the pot and the air surrounding the pot (Figure 8.4.5). Not all input electrical energy is being transformed into the thermal energy of the water.

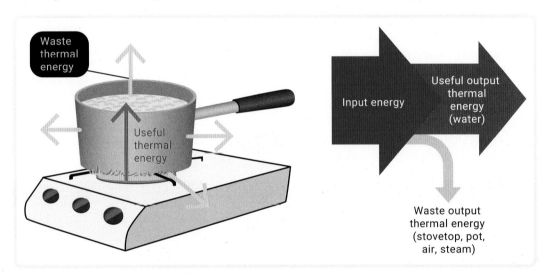

▲ FIGURE 8.4.5   There is waste thermal energy produced when boiling water in a pot.

Thermal energy losses in electricity generation

Electricity generation in a power station involves the following steps (Figure 8.4.6).

1   Chemical potential energy (e.g. usually coal) or nuclear potential energy (e.g. uranium) is converted to thermal energy during a chemical reaction (combustion) or a nuclear reaction.

2   Thermal energy heats water to steam. The steam contains both thermal energy and kinetic energy.

3   Kinetic energy of steam is transferred to kinetic energy of a turbine, which then spins.

4   The spinning turbine produces electrical energy (electricity), using a generator.

9780170472838

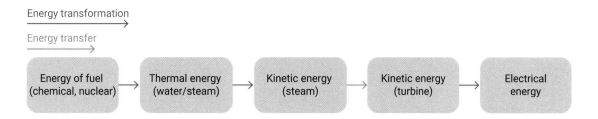

Energy transformation

Energy transfer

| Energy of fuel (chemical, nuclear) | → | Thermal energy (water/steam) | → | Kinetic energy (steam) | → | Kinetic energy (turbine) | → | Electrical energy |

▲ FIGURE 8.4.6   The energy flow diagram for most electricity generation processes

During the energy transfers and transformations between the fuel and the kinetic energy of the turbine, there are many waste energy losses, including:

- thermal energy to the heating of equipment
- thermal energy in steam that escapes or leaves the system after passing the turbine (Figure 8.4.7)
- thermal energy in other exhaust gases that are produced during the reactions of the fuel
- sound energy produced during the movement of mechanical parts.

▲ FIGURE 8.4.7   A lot of thermal energy is lost in electricity production when hot steam is released to the atmosphere.

There is a lot of waste energy when using a fuel to produce electricity. As a result, power generation facilities (power plants) use specially designed processes and equipment to reduce energy losses and increase efficiency.

8.4 LEARNING CHECK

1 **Research** the processes for electricity generation by hydro power, coal and nuclear. **Construct** simple energy flow diagrams for these processes.

2 **Identify** three examples or situations where frictional heat losses occur.

Chapter 8 | Conservation of energy   **281**

8.5 Reducing waste energy

BY THE END OF THIS MODULE, YOU WILL BE ABLE TO:

✓ describe the different forms of waste energy

✓ explain how energy efficiency can be increased and give examples of ways waste energy can be used for other purposes.

Video activity
Improving energy efficiency at home

Other types of waste energy

In Module 8.4, we looked at thermal waste energy. There are also other types of waste energy.

Sound waste energy

Sound energy is a common form of waste energy. There are many energy transformations that are noisy because sound is produced by moving parts and vibrations released into the air. Some common examples are:

- sound made by vehicles when travelling (engine noises, tyres on the road)
- sound produced by the moving parts of machinery
- sound made by a food blender.

Highways often have large noise barriers that absorb sound as vibrations or reflect the sound back to the road.

Electromagnetic radiation waste energy

When light is the intended output of an energy transformation, such as within a light bulb, torch or flame, there are usually other forms of electromagnetic radiation produced. For example, a light bulb can also produce small amounts of ultraviolet (UV) radiation and heat as infrared radiation (Figure 8.5.1).

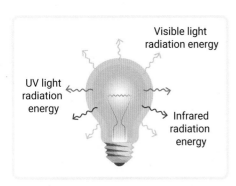

▲ **FIGURE 8.5.1** Waste energy from light globes may include other types of electromagnetic radiation.

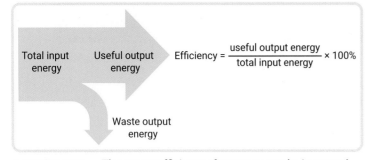

$$\text{Efficiency} = \frac{\text{useful output energy}}{\text{total input energy}} \times 100\%$$

▲ **FIGURE 8.5.2** The energy efficiency of a process can be increased at different stages of the process: by redesigning the process to increase the energy conversion, by reducing waste, and by re-using the waste output energy.

Increasing energy efficiency

There are three common ways to increase the efficiency of an energy transfer or transformation.

- Redesign the energy transfer or transformation process with new technologies that have higher conversions of input energy to useful output energy, such as the development of LED light bulbs.
- Reduce the waste output energy so that less total input energy is required; for example, add insulation to a building.

- Re-use the waste output energy; for example, convert the kinetic energy of steam leaving a kettle to sound energy that indicates the water is boiled.

Redesigning energy transfer and transformation processes often takes a long time and requires large amounts of funding. Reducing or reusing waste energy are generally more practical and affordable ways to increase energy efficiency.

insulation
material used to minimise heat transfer

cogeneration
the process of using waste thermal energy in power generation for another purpose

8.5

Reducing thermal waste energy

In a power plant, water is heated, and the resulting steam moves fast enough to turn a turbine. As the turbine spins, equipment becomes very hot. Heat transfers to the surrounding air and a lot of thermal energy is lost as the hot steam escapes. Minimising heat transfer can reduce the thermal energy losses. This can be done by:

- insulating with materials that reduce heat conduction to equipment or air
- minimising the surface area of containers from which heat loss may occur
- lining containers with reflective surfaces to reduce radiative heat loss.

Houses have **insulation** inside walls, and coverings and double-glazing on windows to reduce heat gain and loss. This increases the efficiency of cooling and heating systems, saving energy and money, and reducing emissions.

▲ FIGURE 8.5.3 Insulation in the walls of buildings reduces the transfer of heat in and out of rooms.

Reusing thermal waste energy

Some power stations are designed to re-use waste thermal energy. **Cogeneration** or combined heat and power generation is the process of using the waste thermal energy for another purpose. Typically, waste steam can be redirected to heat nearby buildings or be used by factories or manufacturing plants in other industrial processes (Figure 8.5.4). Therefore, this process saves energy because less thermal energy is wasted.

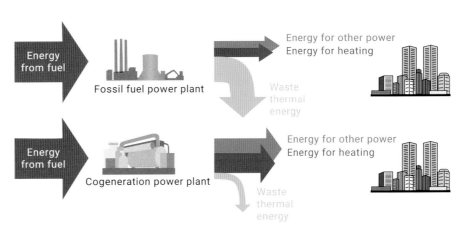

▲ FIGURE 8.5.4 Cogeneration can increase efficiency and reduce fuel consumption by turning waste thermal energy into useful energy.

8.5 **LEARNING CHECK**

1 Consider an electricity production process that supplies 300 MJ of electrical energy and wastes 1000 MJ of thermal energy.

  a **Calculate** the energy efficiency of the process.

  b If 800 MJ of thermal waste energy could be re-used and classified as useful output energy during cogeneration, **calculate** the new energy efficiency of the process.

2 **Justify** why a coal plant fitted with a cogeneration system might be a more environmentally friendly choice than one without.

Chapter 8 | Conservation of energy **283**

8.6 Heat transfer and conservation in First Nations Australians' clothing and bedding

IN THIS MODULE, YOU WILL:

✓ investigate the materials used by First Nations Australians to manufacture clothing and bedding to suit climactic conditions.

▲ **FIGURE 8.6.1** A contemporary possum skin cloak from south-western Victoria

▲ **FIGURE 8.6.2** Possum fur fibres are hollow, which means they trap a layer of air, making fur an excellent insulator.

Clothing worn by First Nations Australians

Many parts of Australia get cold, especially alpine areas, Tasmania and southern parts of mainland Australia. Clothing prevents heat loss from the body. First Nations Australians have long made cloaks from animal furs to control heat loss. In cool climates, clothing was made from the furs of a variety of animals, including wallabies, kangaroos, possums, platypuses and quolls (Figure 8.6.1). The furs are traditionally worn with the fur next to the body, trapping a layer of air as thermal insulation. The Noongar Peoples (south Western Australia) have long manufactured kangaroo skin cloaks called *buka* that traditionally were worn with the fur facing inwards for warmth. Sometimes, furs were rubbed with animal fat to provide a further layer of insulation and waterproofing.

Possum skin cloaks are a culturally important item of clothing, often showing designs that depict connections to Country, kinships, family groups or significant stories. Possum fur is unique because the fibres have a hollow structure. This traps air within the fibre to provide insulation (Figure 8.6.2). Although a possum pelt is smaller than that of many other native mammals such as kangaroos, possum skin provides superior insulation from cold.

Many First Nations Peoples of Australia express their cultural identity through continuing important cultural practices. For example, some First Nations Australians continue to manufacture possum skin cloaks. The Gunditjmara and Yorta Yorta Peoples of Victoria are renowned for the quality of their contemporary cloaks.

Bedding made by First Nations Australians

Mattresses and pillows were manufactured from a variety of natural materials. The Bundjulung Peoples (northern New South Wales) used dried grasses to fill mattresses.

Ngarrindjeri Peoples (South Australia) used dried seaweed to make mattresses. The Dyirbal Peoples (north Queensland) manufactured blankets from sheets of bark from the banana fig. The fibre from the seeds of native kapok trees was widely used as stuffing for pillows. On the Cape York Peninsula, the pods of the large-leaved mangrove were used to provide a soft bedding material. The filling in mattresses, pillows and blankets provides comfort and insulates against weather conditions.

▲ FIGURE 8.6.3   Fibre from the seeds of the native kapok tree is used for stuffing for pillows.

The efficiency of materials in preventing heat loss

☆ ACTIVITY

You need

- 3 small bottles or jars
- hot water
- 3 thermometers
- insulating materials – natural materials such as feathers, wool, fabric, cotton wool, tissues, newspaper, or other material you have chosen
- 3 plastic containers/tubs
- measuring cylinder
- jug

What to do

1 Place a bottle or jar in a plastic container.
2 Fill the surrounding space with one type of insulation material.
3 Repeat steps 1 and 2 for the second insulation material.
4 Place the third bottle or jar in the third container but do not add insulating material.
5 Measure 100 mL of hot water and add it to the first bottle.
6 Immediately record the temperature of the water in the bottle and leave the thermometer in the bottle or jar.
7 Repeat steps 5 and 6 for the other two bottles/jars.
8 Observe the thermometers until one bottle/jar reaches room temperature.
9 When one bottle/jar reaches room temperature, observe and record the temperatures in all bottles/jars.

What do you think?

1 What do your results show?
2 Which material provided the best thermal insulation?
3 Did you conduct a fair test? For example, how did you control the amount of material you used for insulation?
4 How does this model First Nations Australians' knowledge and selection of materials for insulation?

BY THE END OF THIS MODULE, YOU WILL BE ABLE TO:
✓ describe the use of solar energy in Australia and examine the reasons for adoption of solar panels by individuals, industries and communities.

What is solar energy?

Solar energy is the process of converting the electromagnetic radiation from the Sun into electricity or thermal energy. Solar cells or photovoltaic (PV) cells transform light energy to electrical energy. Solar cells are usually grouped together in modules, which are further grouped together to form panels or systems (Figure 8.7.1). Solar panels are placed on roofs or in sunny fields on solar farms where they can transform large amounts of light energy into electricity. Solar energy production is a renewable energy source because it does not involve a fuel that is in limited supply. Using solar panels does not generate greenhouse gases.

Australian researchers have led the way in the development of solar cell technologies. The passivated emitter and rear cell solar cell (PERC) technology was invented by Australian researchers at the University of New South Wales in 1983. This technology remains the basis for more than 85 per cent of all new solar panel modules used globally.

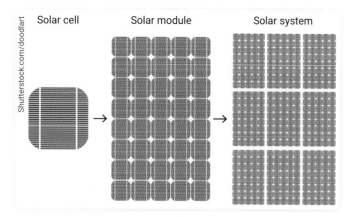

Shutterstock.com/doodlart

Solar cell   Solar module   Solar system

▲ **FIGURE 8.7.1** Solar cells are grouped in modules and within panels or systems for placement on roofs or in solar farms.

Solar use in Australia

Solar energy is the fastest growing electricity generation process in Australia. In 2020–2021, solar cells generated approximately 10 per cent of Australia's total electricity.

Globally, Australia has the highest uptake of residential solar panel use, with one in four homes running solar panel systems (Figure 8.7.2). Homes use solar power to generate electricity and to heat water.

Large-scale solar farms are used for producing hydrogen fuel and energy for large-scale heating and cooling systems (Figure 8.7.3). They also provide electricity for industrial processes and regions. In 2019, there were 24 solar farms in operation or construction in Queensland alone.

Shutterstock.com/Adam Calaitzis

▲ **FIGURE 8.7.2** In 2020, Western Australia's rooftop solar panels generated the same amount of power as the state's biggest coal-powered plant.

The switch to solar

It is no surprise that solar energy is a popular choice in Australia, given we are one of the sunniest countries on the planet. Some of the many reasons people are switching to solar are:

- technological advancement resulting in cheaper and more efficient solar cells
- government investment into large-scale renewable energy production
- financial incentives (discounts and rebates) for people and small businesses to install solar systems
- 'feed-in tariff' schemes where residents and small businesses can sell their additional electricity back to the electricity supply and earn money
- personal choice to minimise environmental impacts.

▲ FIGURE 8.7.3 Solar farms are used in Australia for large-scale electricity production.

8.7 LEARNING CHECK

1 Interview people you know who have had solar panels installed on their home or at their workplace.

  a **Summarise** their reasons for installing solar panels.

  b Make a list of reasons for why other people do not have solar panels installed.

2 **Identify** three reasons why there have been so many solar panels installed in Australia. For each reason, explain why and how it has resulted in more installations.

3 **Create** a flyer that could be dropped in letterboxes around your neighbourhood to encourage people to install solar panels.

4 Look at the two cars in Figure 8.7.4.

  a **Compare** how each car is powered.

  b **Evaluate** which car you think is the best option.

  c **Discuss** the advantages and disadvantages of powering a car with solar energy.

Video activities
Improving solar energy technology

Future of solar energy

▲ FIGURE 8.7.4 Two different ways of powering cars

SCIENCE SKILLS IN FOCUS

IN THIS MODULE, YOU WILL FOCUS ON LEARNING AND IMPROVING THESE SKILLS:

▶ identifying strengths and limitations of primary and secondary evidence and their source

▶ defining credibility and validity

▶ evaluating the credibility and validity of primary and secondary evidence

▶ drawing conclusions from evidence based on interpretation and evaluation.

CREDIBILITY AND VALIDITY OF SECONDARY EVIDENCE

Research or evidence that you use or collect from other researchers is known as **secondary evidence**. This is because someone other than you has conducted the experiment, observation or research. When you use secondary evidence, it is important to think about the credibility and validity of the evidence and its source.

Credibility is the extent to which the evidence can be trusted or considered reliable for use. Credibility is usually based on where the evidence has come from or the method used to collect it. For example, scientific evidence published in a prestigious journal is likely to be more credible than the observations from Bob next door!

The **validity** of the evidence judges how well the evidence answers the specific research question or aim. For example, if a researcher wants to know how plastic pollution is affecting the health of polar bears, but only collects data about penguins, any conclusion they make about their research wouldn't be considered valid. This is because their data does not directly show the relationship between plastic pollution and polar bears.

Some attributes of sources and evidence that can support and question credibility and validity are outlined in Table 8.8.1.

▼ TABLE 8.8.1 Strengths and limitations of different secondary evidence and sources

Strengths	Limitations
Secondary evidence has strong credibility if it is: • written by a reputable university or government department • supported or backed up by the research of other academics or leaders in the field • recent, up-to-date research.	Secondary evidence has limited credibility if it: • is written by a corporate or commercial organisation who may have financial or political reasons for publishing it • has an anonymous author • is out of date or not the most recent research in the field.
Secondary evidence has strong validity if the: • data or ideas are clear and directly related to the research question or aim • methods for research and data collection are clear and scientifically appropriate.	Secondary evidence has limited validity if: • its ideas are only vaguely connected to the research topic • the methods of data collection are unclear.

Evaluating the credibility and validity of the evidence and the source will help you decide whether the conclusions drawn from the evidence are appropriate or not. If you identify limitations with secondary evidence, you may still be able to use it, but you should consider the limitations and discuss them when drawing your own conclusions based on the research or information.

Video
Science skills in a minute: Secondary evidence

Science skills resource
Science skills in practice: Evaluating secondary evidence

INVESTIGATION 1: EVALUATING EVIDENCE TO CHOOSE AN ENERGY-EFFICIENT LIGHTING SOLUTION

To choose the best type of light bulb for lighting a bedroom based on critical evaluation of available research and evidence

Evidence has been collected from a variety of sources. Four pieces of evidence are presented in Table 8.8.2 in the results section.

▼ TABLE 8.8.2 Different types of evidence about different light bulbs and their efficiency

Evidence		Source	Information about the source
1 Quote	'When I switched all the bulbs in our house to halogen bulbs, we saved a lot on our electricity bill and in terms of light in the house, no difference!'	Mark, your friend's dad	• Mark is an engineer. • Mark is a very nice man. • Mark is a close friend and always gives good advice.
2 Quantitative data	**Light bulb** / **Power required for 500 lumens of light (W)** LED / 5–8 CFL / 7–9 Halogen / 28 Incandescent / 40 © Commonwealth of Australia 2019/ energyrating.gov.au (CC BY 3.0 AU)	Australian Government website	• The initiative is led by the Australian Government, state and territory governments and New Zealand Government. • It is managed by the Greenhouse Energy Minimum Standards Regulator and the Energy Efficiency Advisory Team made up of government officials. These officials undertake activities to improve the energy efficiency of appliances and equipment, including energy rating labelling, setting minimum energy performance standards and education and training programs.
3 Graph	© Commonwealth of Australia 2019/ energyrating.gov.au (CC BY 3.0 AU)		
4 Quantitative data		Aussie Hardware company website	• Aussie Hardware is a leading Australian retailer of home and lifestyle products. • The retail company also owns a number of other commercial brands and companies. • Digital branding for the company includes online columns and blog posts for advice, inspiration and product reviews.

1 Analyse the pieces of evidence separately to identify what each piece can tell us about the best choice for an energy-efficient light bulb.

2 Construct a table such as Table 8.8.3 to identify strengths and limitations associated with each piece of evidence.

▼ TABLE 8.8.3

Evidence	Strengths	Limitations
1		
2		
3		
4		

3 Rank each piece of evidence in order of least to most credible. Justify your choice using the evidence in Table 8.8.2.

4 Rank each piece of evidence in order of least to most valid. Justify your choice using evidence in Table 8.8.2.

CONCLUSION

Draw a conclusion about which type of light bulb you would choose to use in your bedroom to save energy. Justify your choice by referring to your responses in the analysis and evaluation section.

INVESTIGATION 2: MEASURING THE EFFICIENCY OF A BALL BOUNCE

AIM

To measure and compare the efficiency of a ball bounce for two balls made of different materials

MATERIALS

☑ 2 different balls that will bounce (e.g. ping pong ball, tennis ball, bouncy ball)
☑ metre ruler
☑ putty-like adhesive
☑ video recording device
☑ electronic scales

METHOD

1 Hold a metre ruler against a flat wall so that it measures from the ground up.

2 Secure the ruler in place with the adhesive. Alternatively, you can hold the ruler in place against the wall.

3 Weigh each ball on the electronic scales and record their masses.

4 Hold the first ball so that the bottom of the ball is at the 1.0 m mark.

5 Do a test bounce to confirm approximately where the ball will bounce to when dropped from this height.

6 Set up the video recording device so that it will record the area where the ball will bounce to at a level angle.

7 Hold the ball again so that the bottom of the ball is at the 1.0 m mark.

8 Start recording and drop the ball. Allow it to bounce once before stopping it and the recording.

9 Use the recording to measure the highest point reached by the bottom of the ball.

10 Record this height as the bounce height for the 1.0 m drop height.

11 Repeat steps 6–10 for two more trials.

12 Calculate an average bounce height for the 1.0 m drop height.

13 Repeat steps 5–11 for drop heights of 0.8 m, 0.6 m, 0.4 m and 0.2 m.

14 Repeat steps 3–13 for the other ball.

RESULTS

Record your results in a table like Table 8.8.4.

▼ **TABLE 8.8.4** Experimental results for the bounce height of each ball at different drop heights

Ball type	Mass (kg)	Drop height (m)	Bounce height (m)			
			Trial 1	Trial 2	Trial 3	Average
		1.0				
		0.8				
		0.6				
		0.4				
		0.2				
		1.0				
		0.8				
		0.6				
		0.4				
		0.2				

1 Use your table to calculate the input energy for each drop height, using the formula:

$$E_{\text{Pdrop}} = mgh_{drop}$$

where E_{Pdrop} is the gravitational potential energy at the drop height (J), m is the mass of the ball (kg), g is the gravitational acceleration value of 9.8 m/s^2, and h_{drop} is the drop height of the ball (m).

For example, for a ball that weighs 0.250 kg and a drop height of 0.8 m:

$h_{drop} = 0.8$ m

$g = 9.8$ m/s^2

$m = 0.25$ kg

$E_{Pdrop} = mgh_{drop}$
$= 0.25 \times 9.8 \times 0.8$
$= 1.96$ J

2 Calculate the useful output energy for each drop height using the formula:

$$E_{\text{Pbounce}} = mgh_{bounce}$$

where E_{Pbounce} is the gravitational potential energy at the bounce height (J), m is the mass of the ball (kg), g is the gravitational acceleration value of 9.8 m/s^2, and h_{bounce} is the bounce height of the ball (m).

3 Calculate the efficiency of the ball bounce for each drop height and record this in Table 8.8.5.

4   Calculate the average efficiency of a ball bounce for each type of ball and record this in Table 8.8.5.

▼ TABLE 8.8.5 Processed data table for ball bounce efficiency

Ball type	Drop height (m)	Input energy (J)	Useful output energy (J)	Efficiency (%)
	1.0			
	0.8			
	0.6			
	0.4			
	0.2			
				Average
	1.0			
	0.8			
	0.6			
	0.4			
	0.2			
				Average

ANALYSIS AND EVALUATION

1   **Construct** a scatter graph of bounce height (m) (*y*-axis) versus drop height (m) (*x*-axis) for both balls. Include both sets of data on the one plot and include a legend to identify the different balls.

2   **Construct** a column graph to compare the average efficiency of the balls.

3   **Identify** the trend shown in the scatter plot from step 1.

4   **Compare** the average bounce efficiencies of the balls.

5   **Discuss** the variability of the trials used to calculate the average bounce height.

6   **Explain** how possible measurement errors may have affected the data recorded.

7   **Identify** other experimental errors from this experiment and describe how they may have affected the data collected.

8   **Suggest** two ways in which this experiment could be improved.

CONCLUSION

Draw a conclusion that directly responds to the aim of the experiment.

REMEMBERING

1 **Recall** the law of conservation of energy.

2 **Recall** the formula for calculating energy efficiency.

3 **Describe** the process of cogeneration.

4 **Define**:

a waste energy output.

b useful energy output.

c Sankey diagram.

UNDERSTANDING

5 **Define** the following terms and give an example for each.

a Energy transfer

b Energy transformation

6 **Describe** the concept of energy efficiency and what it measures.

7 **Explain** how output energy is classified as useful or waste.

APPLYING

8 A ball thrown into the air starts with 450 J of kinetic energy. If it loses 15 J through frictional heat losses on its way up, **determine** how much gravitational potential energy it has at its maximum height before returning to the ground.

9 **Explain** how a sound barrier next to a freeway relates to wasted energy.

10 **Calculate** the efficiency of a process that has 200 J of input energy and 40 J of useful output energy.

11 Ceiling fans convert electrical energy into kinetic energy to move air. **Calculate** the kinetic energy produced per second from a ceiling fan that uses 400 J/s and is 80 per cent efficient.

12 **Calculate** how much waste energy is produced by a fridge that uses 700 000 J of electrical energy and is only 60 per cent efficient.

13 **Explain** why bicycle brakes get hot.

14 **Compare** the advantages and disadvantages of possum and kangaroo furs as materials used to make cloaks.

ANALYSING

15 If you walk to the shops, you might need about 200 kJ of chemical energy from your food. If you drove a car there, the chemical energy from the petrol needed would be much more. **Explain** the difference in terms of useful output energy and waste output energy.

16 The following diagram is taken from a video explaining how to use the star rating system in the Energy Rating Label scheme. **Interpret** the main message from this diagram and identify one key limitation of the system. **Explain** why this limitation exists.

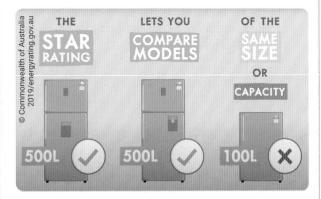

17 When a car is moving, it transforms chemical energy into kinetic energy. A lot of waste heat is produced. **Construct** a Sankey diagram to illustrate this. The Sankey diagram does not need to be to scale because not enough information has been given to do so.

EVALUATING

18 **Explain** why energy efficiency is important when choosing appliances.

19 Consider two power plants: one that has cogeneration design features and one that does not. **Construct** a table to **evaluate** the advantages and disadvantages of each design.

20 **Compare** the energy transformations when using a hairdryer and when hitting a ball with a bat. Use Sankey diagrams to identify similarities and differences in these two processes.

CREATING

21 **Construct** energy flow diagrams for two hot water systems: one that heats water via wind-generated electricity and one that heats water via solar-generated electricity.

22 **Construct** a Sankey diagram to scale for the following situation.

A slingshot transforms elastic potential energy into kinetic energy to send objects flying through the air. A slingshot holding a small pebble is stretched backwards and stores 40 J of elastic potential energy. The pebble is projected upwards and the slingshot is released, transferring 38 J of kinetic energy into the pebble. During its flight, the conversion of kinetic energy to gravitational potential energy has 90 per cent efficiency.

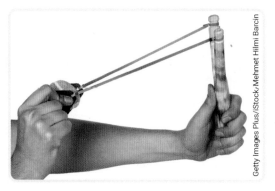

23 **Construct** a concept map for the key words for the chapter. Join words that are related and annotate each connecting line with a justification for how they are connected.

BIG SCIENCE CHALLENGE PROJECT #8

1 **Connect what you've learned**

In this chapter, you have learned about the law of conservation of energy, energy efficiency, waste energy and many applications of these ideas. Create a mind map to show how the main ideas you have learned about are connected.

2 **Check your thinking**

Compare the methods of electricity production for solar, wind, coal, nuclear and hydroelectricity by identifying similarities and differences between them.

Shutterstock.com/Wirestock Creators

3 **Make an action plan**

Consider one problem with sourcing energy in Australia now. For this problem, complete the following.

- Describe the problem.

- Suggest a possible solution or action towards a solution that could be taken.

- Identify who needs to be involved for this solution to be actioned.

- Describe the effects this solution may have.

- Suggest ways that you can be involved in this solution.

4 **Communicate**

Draft a letter to your local member of parliament (MP) that includes the following.

- What you know about sourcing energy for electricity in our country

- How you think we should be sourcing our energy and what actions the government should take

- Questions you might have about energy production in Australia and where it comes from or could be coming from

Dr Steven Chu is a US physicist who worked with President Obama to address the climate crisis. He wrote a famous letter to his government in which he made the following quote; 'The Stone Age did not end because we ran out of stones; we transitioned to better solutions.'
Write a paragraph to discuss what you think Steven Chu meant when he said this and what you think this letter might have been about.

Glossary

absolute zero the lowest temperature possible (−273°C) at which matter is considered to have no thermal energy

absorption the process of taking in radiation and converting it to other types of energy

acid a substance that can donate a hydrogen ion when in solution

adenosine triphosphate (ATP) a useable form of energy in living things

aerosol fine droplets of saliva or mucus containing pathogens

alpha particle a helium nucleus (two protons and two neutrons) emitted when unstable larger nuclei decay

amplitude the maximum distance that a particle moves when a mechanical wave passes through it

anther the section of the male part of a flower (stamen) that produces pollen grains

antibody a Y-shaped protein that binds to one type of antigen

antigen a substance on a pathogen that stimulates the adaptive immune response

aqueous a solution containing a substance dissolved in water

artificial immunity the protection obtained from the production of antibodies after receiving a vaccine or antibodies found in medicine

artificial isotope an isotope of an element produced in a nuclear reactor or particle accelerator

asexual reproduction a method of reproduction that involves one individual producing an identical copy of itself

atmosphere all of the gases surrounding Earth's surface

atom the fundamental particle of matter; made up of protons, neutrons and electrons

atomic number the number of protons in the nucleus, which is the same for every atom of the same element; symbol Z

axon a long, thin fibre that carries electrical impulses from the cell body of a neuron towards the next neuron

B cell a type of white blood cell involved in the humoral (antibody)-mediated immune response

balanced chemical equation a chemical equation in which there are the same number of atoms of each element on each side

base a substance that can produce hydroxide, oxide or carbonate ions in solution

beta particle a negatively charged particle identical to an electron, emitted when an unstable nucleus decays

binary fission a method of asexual reproduction in bacteria in which a parent cell splits into two identical daughter cells

bioethanol ethanol produced from crops through the process of fermentation

biogeochemical cycle a model describing how chemical substances are transformed and stored in Earth's spheres

biosphere all the parts of Earth and the atmosphere that support life, including all living things

birth the process in which the foetus is pushed out of the uterus and leaves the mother's body

blood glucose the amount of glucose in blood

booster an additional dose of a vaccine, designed to provide a higher level of protection when antibodies have decreased

brain a very complex organ, the coordinating centre of the CNS

brain stem consists of the midbrain, the pons and the medulla oblongata; controls automatic and involuntary activities such as breathing, heart rate, digestion and vomiting

budding a method of asexual reproduction in which a bud (a growth on the parent body) forms, grows and then separates or spreads

bulbourethral gland one of two glands located on both sides of the urethra, below the prostate gland, that secrete a lubricating fluid

capsid the protein coating that protects the genetic material inside a virus

carbon an element in all organisms; its atoms can form four bonds that store energy

carbon-14 dating the use of the carbon-14 decay process to determine the age of materials containing carbon

carbon cycle the transfer and transformation of carbon through Earth's spheres

carbon footprint the amount of carbon dioxide emitted by a person, organisation or event

carbon sequestration the process of capturing and storing atmospheric carbon dioxide

carbon sink a type of reservoir where more carbon is absorbed than is released

carpel the female reproductive part of a flower, consisting of the stigma, style, ovary and ovules

cathode ray a beam of electrons produced by a cathode ray tube

cell body the part of a neuron that contains the cytoplasm, including the nucleus

cell-mediated immunity the specific immune response prompted by T cells on infected host cells

cellular respiration the process in which producers and consumers break down organic compounds (e.g. glucose) to produce ATP (energy)

central nervous system (CNS) the brain and spinal cord

cerebellum a smaller section at the back of the brain; controls movement and balance

cerebrum the largest part of the brain; controls movement, learning, emotion, memory and perception

cervix a muscular opening between the vagina and uterus

chemical change a change that results in the formation of a new substance in a chemical reaction

chemical equation a way of presenting information about chemical reactions

chemical formula symbols and numbers used to represent the composition of a chemical

chemical reaction a process involving the rearrangement of atoms in reactants to form new chemical products, through a chemical change

chlorophyll a chemical pigment, usually green, found in chloroplasts

chloroplast a cell organelle in producers that absorbs the light energy for photosynthesis

cilia a hair-like extension of mucous membranes that pushes pathogens up and out of the body

9780170472838

clean hydrogen hydrogen that is produced with low or no carbon emissions

coal a solid form of a fossil fuel

coefficient a number placed in front of a chemical formula to balance an equation

cogeneration the process of using waste thermal energy in power generation for another purpose

combustion the burning of substances, such as fossil fuels, in oxygen to produce heat and light

complete combustion combustion occurring in excess oxygen, producing carbon dioxide and water

compression a high-pressure region of a longitudinal mechanical wave where particles are pushed close together

condense to change state from a gas to a liquid

conditions factors that affect a chemical reaction, such as heat or pressure

conduction the transfer of heat between particles by direct contact

consumption the process of a consumer feeding on a producer or another consumer, resulting in a transfer of energy and carbon

convection the transfer of heat within fluids (liquid or gas) by particle movement

coordinating centre an organ or tissue that receives and processes information from receptor cells and coordinates a response

corpus callosum a collection of axons that extend across the two hemispheres of the brain that compares and combines the sensory inputs from the left and right sides of the body

corrosion a chemical process that often happens to metals exposed to oxygen; also known as rusting

crest the highest point on a transverse wave

cross-pollination the transfer of pollen grains from one flower to a flower on a different plant

crossing over swapping of small amounts of maternal and paternal genetic material; happens before random assortment

crude oil an unrefined liquid form of a fossil fuel

cycle continuous, connected and repeated processes that move and transform matter through Earth's spheres

cytosol the liquid substance inside a cell that holds organelles in place

 D

decomposition breaking down complex organic matter into simple inorganic molecules in a form that can be taken up again by plants

dendrite a branching network at the end of a neuron that receives information from other neurons

density a measure of mass per unit volume or how compact mass is within a certain space

dispersion the separation of light into its different colours

DNA deoxyribonucleic acid, the molecule that makes up the genetic material inside cells and in some viruses

 E

effector a muscle or gland that receives a message from the coordinating centre and carries out a response

electromagnetic radiation energy that travels in electromagnetic waves

electromagnetic spectrum the range of electromagnetic waves arranged in sequential order from high-energy gamma rays to low-energy radio waves

electromagnetic wave a wave that transfers energy through space by electric and magnetic fields

electron a negatively charged particle in an atom, which moves in space around the nucleus

electrostatic force a force of attraction between oppositely charged particles

element a substance consisting only of atoms that have the same number of protons in their nuclei

endocrine system a network of glands that secrete hormones into the bloodstream to be transported to target cells

endometriosis a painful disorder experienced by some women when tissue similar to uterus lining grows on the outside of the uterus

endothermic a chemical reaction that absorbs energy

energy the ability to do work; measured in joules (J), kilojoules (kJ) or megajoules (MJ)

energy efficiency a measure of how much input energy is converted to useful output energy, often stated as a percentage

energy flow diagram a visual representation of energy movement using arrows and boxes

energy transfer the movement of energy without changing its type

energy transformation a movement of energy where it changes from one form into another

enhanced greenhouse effect the increased warming of the atmosphere and Earth's surface due to greenhouse gases produced by human activities

epididymis a tube that sits on the back of each testicle; stores sperm

ethanol a chemical containing carbon, hydrogen and oxygen that can be used as a fuel

ethics moral considerations that govern decisions or actions that can affect people or other organisms

evaporative cooling the cooling effect that occurs when water evaporates

exothermic a chemical reaction that releases energy

 F

fallopian tube a long narrow tube that connects one ovary to the uterus and is the site of fertilisation; sometimes called the oviduct

fermentation the chemical reaction that uses yeast to convert glucose to ethanol

fertilisation the process in which a male reproductive cell and a female reproductive cell (gametes) from each parent join to make the offspring's first cell, a zygote

fever a non-specific response to infection where the body temperature exceeds the normal 37°C

filament a tall structure that elevates the anther so that it sticks out of the flower, making it visible to pollinators

fossil fuel a substance containing hydrocarbons that is used as an energy source; takes millions of years to form from the remains of dead plants and animals

fragmentation a method of asexual reproduction in which a body part breaks off from a parent body and then develops into a new offspring

frequency the number of waves produced, or passing a point per unit of time; measured in hertz (Hz)

friction a resistance force that results when two surfaces rub against one another

frictional heat loss waste thermal energy produced as a result of the vibration of particles when moving past each other in a process where thermal energy production is not intended or useful

gamete a sex cell of a sexually reproducing organism

gamma radiation high-energy electromagnetic energy emitted when an unstable nucleus decays

geosphere the rocks, minerals and landforms of Earth's surface and its interior

gestation the period between fertilisation and birth, during which a foetus (unborn baby) develops

gland a tissue that releases hormones

glucagon a hormone secreted by the pancreas; breaks down glycogen into glucose

glycogen a store of glucose in the liver and muscles

gonad a male or female reproductive organ such as the testicles and ovaries

greenhouse effect a natural process that traps energy within the atmosphere, causing Earth's surface temperature to increase

greenhouse gas a gas in the atmosphere that can trap heat and affect global surface temperatures

half-life the time it takes for half of the nuclei in a sample of an isotope to decay

heat thermal energy that is being transferred between different places or particles

hemisphere one of the two symmetrical halves of the brain

herd immunity a level of protection gained when a certain percentage of a community is immune to an infectious disease, which breaks the chain of transmission

homeostasis the maintenance of a constant internal environment necessary for survival

hormone a chemical messenger

host an organism infected with a pathogen

humoral immunity the specific immune response prompted by B cells, and antibodies specific to the antigen, on pathogens found outside of host cells

hydrocarbon a substance containing the elements hydrogen and carbon

hydrogen ion a hydrogen atom that has lost its electron; represented by H^+

hydrosphere the component of Earth's system that consists of all the solid, liquid and gaseous forms of water

hypothalamus a pea-sized structure of the brain; plays a major role in homeostasis

in vitro fertilisation (IVF) a complex series of procedures that involve the fertilisation of eggs in a laboratory and implanting the embryo (fertilised egg) in the uterus

incident wave a wave of light that is approaching a surface

incomplete combustion combustion occurring in limited oxygen, producing carbon, carbon monoxide and water

indicator a chemical that changes colour in acidic and basic solutions

infertile a term used when a female has been unable to get pregnant after one year of trying

infiltration movement (soaking) of water into the surface of Earth

inflammation a response triggered by damaged cells in the body or a pathogen; characterised by redness, swelling, heat and pain

infrared radiation electromagnetic radiation that transfers heat through space

insulating reducing the transfer/loss of heat

insulation material used to minimise heat transfer

insulator a material that restricts heat transfer by conduction

insulin a hormone secreted by the pancreas; controls how much glucose is in the blood

interneuron a short neuron that sends nerve impulses between sensory and motor neurons within the CNS

ion a charged particle that forms when an atom gains or loses electrons

ionising power the ability of radiation to change the structure of materials it passes through

isotopes atoms of an element with the same number of protons but different numbers of neutrons

law of conservation of energy when energy is transferred or transformed, the total amount of energy remains the same

law of conservation of mass the total mass of reactants and products in a chemical reaction is equal

longitudinal wave a mechanical wave in which particles oscillate in the same direction in which the wave is travelling

lymphocyte a type of white blood cell involved in specific immune responses

lysosome a cell organelle that contains digestive enzymes

lysozyme an enzyme that breaks down pathogens

mass number the total number of protons and neutrons in the nucleus of an atom; symbol A

mechanical wave a wave that passes energy through space between particles

medium the matter or substance through which a wave passes (made from particles)

meiosis cell division that produces four non-identical gametes with half the genetic material of the parent cell

memory cell a mature B cell or T cell that 'remembers' information about a specific antigen

metal oxide a chemical substance made up of a metal and oxygen

mitochondrion an organelle where aerobic cellular respiration occurs and produces a usable form of energy

mitosis cell division and a form of asexual reproduction in some simple organisms

model a conceptual, physical or mathematical representation of an idea, phenomenon or process

molecule two or more non-metal atoms bonded together

motor neuron a neuron that sends electrical impulses from the CNS to effectors

mucus thick, sticky liquid in our airways that captures pathogens

mutation a permanent change in the DNA in the cell of an organism

myelin sheath a protective coat around an axon that increases the speed of nerve impulses

natural gas a gaseous form of a fossil fuel; usually methane, propane or butane

natural immunity the protection obtained from the production of antibodies after exposure to an infectious disease

natural isotope an isotope of an element found in nature

nectar a sugary liquid in the middle of a flower, produced by the flower, which attracts pollinators

negative feedback a mechanism in which a response reduces a condition or factor back to its optimal range

nerve a collection of fibres, surrounded by a protective coat, that transmit messages as nerve impulses to and from the CNS

nerve fibre the section of a nerve cell that carries nerve impulses away from the cell body

nerve impulse an electrical message that is transmitted along nerves to and from the CNS

neuron a specialised cell that can transmit nerve impulses; also known as a nerve cell

neurotransmitter a chemical signal that delivers a message that was in the form of an electrical impulse

neutron a particle in the nucleus of an atom that does not have an electrical charge

non-renewable a substance that cannot be easily replaced at the same rate it is consumed

offspring a new organism produced by asexual or sexual reproduction

organic compound a compound containing carbon, usually bonded to other carbon and hydrogen atoms

oscillation movement back and forth in a regular rhythm or pattern

ovary (animal) the female gonad where the ova (female gametes/eggs) are produced

ovary (plant) the structure at the base of a flower that encloses the ovules, which contain the female gametes

ovulation the release of an ova from an ovary

ovules structure inside the ovary that makes and contains female gametes

ovum a female gamete, or egg, produced in the ovary of a female

oxidation a chemical process in which oxygen is added to a substance

pandemic a disease that has spread rapidly across the world, having grown from an epidemic

parthenogenesis a method of asexual reproduction in which an unfertilised egg matures and develops into offspring without fertilisation by sperm

particle accelerator a machine that enables high-speed, high-energy collisions between atoms to produce artificial isotopes and new elements

particle a small, localised piece of matter, usually modelled to be spherical

pathogen an organism or virus that can cause an infectious disease

penetrating power the ability of radiation to pass through materials

penis a sex organ that inserts into the vagina to transfer sperm into the female reproductive system

percolation the movement of water through soil

period the time taken for one full wave to pass a particular point

peripheral nervous system (PNS) the network of nerves outside the brain and spinal cord

petal the colourful part of a flower that attracts pollinators

pH a measure of how acidic or basic a substance is

phagocyte a specialised white blood cell that can engulf and destroy pathogens

phagocytosis a process in which blood cells called phagocytes engulf and destroy a pathogen

phenomenon something that is observed to exist or occur (plural: phenomena)

photosynthesis the process in which producers use light energy to convert carbon dioxide and water into sugars

physical change a change of state or appearance of a substance or an object without a change in its chemical composition

pitch the degree of frequency (high or low) of a sound or note

pituitary gland a gland that produces and secretes several hormones; is connected to, and is controlled by, the hypothalamus

pollen grains that produce and hold male gametes (sperm) on the anther of a flower

pollen tube a tube that develops when a male gamete (inside pollen) lands on a stigma

pollination the transfer of pollen from a male anther to a female stigma

pollinator an animal that transports pollen from the male anther to the female stigma for pollination

population a group of individuals of the same species that live in the same location

positive feedback a mechanism in which a response reinforces the condition or factor until an end point

precipitation liquid or solid water in the form of rain, snow, sleet or hail

producer an organism such as a plant that converts light energy into chemical energy

product a chemical formed in a chemical reaction

prostate gland a gland that produces fluids found in semen mix, located between the bladder and the penis

proton a positively charged particle in the nucleus of an atom

radiation a stream of particles and/or energy from a radioactive source

radioactive decay the spontaneous disintegration of certain atomic nuclei accompanied by the emission of alpha particles, beta particles or gamma radiation

radioactive isotope (radioisotope) an isotope that is unstable and undergoes radioactive decay to become more stable

radiotherapy the treatment of cancer by radiation

random assortment random recombining of large amounts of maternal and paternal genetic material; happens after crossing over

rarefaction the low-pressure region of a longitudinal mechanical wave where particles are spread apart

reactant an initial chemical in a chemical reaction

receptor a specialised cell that detects a stimulus; may be internal or external

reflected wave a wave of light that has been reflected off a surface

reflection the process of a wave bouncing off a surface and returning in the opposite direction through the same medium

refraction the transmission of a wave from one medium into another with a resulting change of direction

reservoir (carbon) a long-term storage area of carbon, such as the atmosphere, the biosphere, oceans and sediments

reservoir (infection) a source of infection such as a habitat or an organism where a pathogen can replicate or survive for long periods

response the action of an effector that reverses a stimulus

RNA ribonucleic acid, a molecule similar to DNA; the genetic material found in many viruses

Sankey diagram a type of flow chart that uses arrows of various sizes to indicate the amount of energy transferring or transforming in a system

scrotum skin in the shape of a sac that wraps around the testes to keep sperm cooler than normal body temperature

seed the structure that forms from the ovule after it is fertilised, usually found in fruit

self-pollination the transfer of pollen within the same flower or plant

semen a mix of sperm and fluids that provide nutrition and lubrication for sperm

seminal vesicle one of two sac-like glands that secrete fluids that are the main component of semen

sensory neuron a neuron that sends electrical impulses from receptors to the CNS

sepal one of many structures that cover the flower (bud) as it forms

sexual reproduction a method of reproduction that involves two parents producing offspring that are not identical to the parent or each other (except twins)

shell (electron shell) an energy level around the nucleus of an atom containing electrons of the same energy

spawning the release of male and female gametes into the environment (usually aquatic), during which sperm randomly fertilise eggs

spectrum a range or a scale

speed a measure of how fast a wave is travelling; a measure of the distance covered by the wave per unit of time; measured in metres per second (m/s)

sperm a male gamete produced in the testes of male animals, or inside pollen grains of plants

stamen the male reproductive part of a flower, consisting of a filament and anther

stigma the sticky tip of the carpel where pollination occurs

stimulus a mechanism that causes an increase or decrease in a factor above or below the optimal range; causes a movement away from normal

strong nuclear force a force of attraction between particles in the nucleus of an atom

style a long stalk that connects the ovary to the stigma

subatomic particle a particle inside an atom, such as a proton, a neutron or an electron

supercomputer a powerful computer with multiple processors that run in parallel, significantly faster than a single computer

synapse the gap between two neurons

system a set of interacting processes and components

T cell a type of white blood cell involved in the cell-mediated immune response

temperature a measurement of the kinetic energy within a substance that lets us know how hot or cold something is

testes the male gonads that produce sperm (male gametes/sex cells)

thermal energy the energy associated with a body resulting from the kinetic energy and potential energy of particles

tolerance range the range of a particular condition inside the body that an organism can survive

transmission the transfer of a pathogen from one organism or reservoir to a new host

transverse wave a mechanical wave in which particles oscillate at right angles to the direction in which the wave is travelling

trough the lowest point on a transverse wave

uranium–lead dating the use of either the uranium-238 or uranium-235 decay process to determine the age of rocks

urethra a tube that transports urine and semen, extending from the bladder to the tip of the penis

useful output energy the output energy of a process or action that is intended and useful

uterus the organ where a fertilised egg implants and develops into an embryo and foetus

vaccination administration of a vaccine to provide protection from a specific disease

vaccine a substance that stimulates a response in the immune system and produces antibodies

vacuum a space where there is no matter (no particles)

vagina a muscular opening that leads from the external environment to the cervix

vas deferens a long, muscular tube that transports sperm from the epididymis to the urethra

vector an organism that transmits a pathogen from an animal or a plant to another animal or plant

9780170472838

vegetative reproduction a method of asexual reproduction in which new plants form from vegetative parts of the original plant, such as the leaves, stems and roots

viral envelope the outer layer of many viruses

volume the measure of how loud a sound is; measured in decibels (dB)

waste output energy the output energy of a process or an action that is not intended and not useful for the main purpose of the process or action

wave a disturbance of energy that moves through matter or space

wave equation a mathematical relationship between the speed, wavelength and frequency of a wave

wavelength the length of one full wave; the distance between two crests or two troughs; measured in metres (m)

white blood cell a specialised cell in the blood that defends the body against disease

zygote the cell that results from the joining of a sperm and an egg cell at the completion of fertilisation

Index

9780170472838

Extra credits